Innovations and the Environment

Yoram Krozer

Innovations and
the Environment

 Springer

Yoram Krozer, PhD
Cartesius Institute
Institute for Sustainable Innovations
  of the Netherlands' Technical Universities
Zuidergrachtswal 3
8933 AD Leeuwarden
The Netherlands

ISBN 978-1-84996-750-1   e-ISBN 978-1-84800-197-8

DOI 10.1007/978-1-84800-197-8

British Library Cataloguing in Publication Data
Krozer, Yoram
  Innovations and the environment
  1. Technological innovations - Environmental aspects
  2. Environmental protection - Technological innovations
  3. Technological innovations - Economic aspects
  4. Sustainable development
  I. Title
  608

Cover design: eStudio Calamar S.L., Girona, Spain

Printed on acid-free paper

9 8 7 6 5 4 3 2 1

springer.com

Preface

This book Innovations and the Environment aims to review the state of knowledge and to provide an empirical basis for the development and use of new technologies, products and services to reduce the impact of human activities on environmental quality. This is at the core of the debate on economics and the environment. The message is optimistic. Advances in thinking and actions make it possible to maintain high environmental quality at reasonable costs, sometimes even having a positive effect on corporate results. The progress is by no means automatic. Risk taking by entrepreneurs, managers and policy makers is necessary.

The idea to write a book about innovations and the environment was born during discussions with Professor Han Brezet who was convinced of the importance of this subject for engineering students. This is because he has found many stories about the successes and failures of environmental technologies and even more on the impact of technologies on environmental qualities, but no reviews on technology and the mindset of entrepreneurship to explain the forces that drive innovations for sustainable development.

Profitability is this book's starting point because it is essential for any innovator. The basic question is how to achieve profitable innovations for improvements in environmental quality, or as my daughter Mira put it: How can nagging about the environment pay? The material included is largely based on the experiences of companies and authorities in the Netherlands during the 1980s and 1990s that have been global trailblazers in innovating for the environment. The book is a result of my entrepreneurial activities at the Institute of Applied Environmental Economics. Partners and colleagues at the Institute provided a lot of support during its writing. In particular, I am grateful to Jochum Jantzen for the use of so much empirical material on the cost of technologies. Petra Doelman helped me a lot with her experience in business development at starting innovative companies CES and Avanti, and in Philips. Most of all, it was exciting to co-operate with brilliant people working in many companies. I will mention only a few by name, the creators of environmental management who are unfortunately no longer with us: Cees van Leeuwen, environmental manager for the Unilever companies; Jan de Haas, quality manager for retailer Vroom & Dreesman; Cornelis Betlem, business developer at Ahrend, Bram van de Drift of Philips and Coos

Veldman, general director at Wagenborg Shipping. These men have moved environmental management towards the highest standards of social responsibility.

The book was finished during my academic work at the Cartesius Institute. In particular, Professor Andries Nentjes a founding father of environmental economics in Europe helped me avoid many scientific traps through his strong guidance. I am indebted to Simon Tijsma at the provincial authority of Friesland for his valuable comments on conclusions regarding the realities and peculiarities of policy making. Lisa Hayes was extremely helpful with language and Satish Kumar Beella with editing.

This book is dedicated to my mother, Raisa Grigorievna.

Contents

1

Introduction

Economists widely recognize the importance of the natural environment for welfare and productivity. The founder of modern economic theory, Adam Smith, wrote in 1776: "The beauty of the country besides, the pleasure of a country life, the tranquillity of mind which it promises, and wherever the injustice of human laws does not disturb it, the independency which it really affords, have charms that more or less attract everybody; and as to cultivate the ground was the original destination of man, so in every stage of his existence he seems to retain a predilection for this primitive employment" (Smith, 1979, page 481).

In the 19th century, the pastoral community that had prevailed so far was threatened by the industrial revolution. Hence, 70 years after Adam Smith, John Stuart Mill remarked sadly in his economic masterwork: "If the earth must lose that great portion of its pleasantness which it owes to things that the unlimited increase of wealth and population would extirpate from it, for the mere purpose of enabling it to support a larger, but not a better or happier population, I sincerely hope, for the sake of prosperity, that they will be content to be stationary, long before the necessity compels them to it" (Mill, 1970, page 116). Mill and many other scholars from his generation argued that the population growth occurring due to better public health jeopardized the environmental quality needed for welfare growth. Seventy-five years later, in 1920, Pigou emphasized that it was industrialization that limited welfare growth because the side effects of industry such as pollution caused social hazards and damaged the environment. He pinpointed a poor legal framework in market economies as the main cause of problems and advocated a legal framework to prevent these side effects caused by industrialization (Pigou, 1920). Fifty years later, in his reflections on 150 years of debate on "economy versus environment", Rosenberg (1973, 1975a) used historical material to illustrate that the question of whether to limit production growth and even reduce it, or to grow and prevent environmental damage through innovation is answered in favor of growth with changes towards less polluting solutions. These changes are by no means automatic. He argued that innovations to prevent degradation of environmental qualities were not spontaneous entrepreneurial actions but rather efforts triggered by the authorities in view of scarce resources and environmental harm. Today, policies to control the side effects of economic development and create conditions for clean technological progress are widely accepted.

Good environmental qualities are increasingly demanded, such as access to space, energy, minerals, amenities, diverse nature, heritage and landscape and in particular, clean air, water and soil that make a pollution-free environment. Pollution is a focal environmental issue that is addressed in this book. Pollution is caused by emissions produced from millions of sources involved in industries, agriculture, households, traffic and other activities. The pressing issue is how to reduce emissions to the level that enables healthy living among all people and nature, including vulnerable children and sensitive, biological organisms. Emissions reduction is necessary for productive activities such as good soil for agriculture and for basic consumptive needs such as healthy food, safe drinking water and sound hygienic conditions. Progressively reducing emissions must be achieved to counter the increase of pollution related to production growth and to prevent the impacts of pollution on climate, biodiversity and human lives such as the greenhouse effect, contamination, epidemics and others. Studies on the environment suggest that most emissions must decrease to one-tenth or less of the present level in the next 15–20 years in order to maintain the availability of environmental qualities. We do not debate the figure but take for granted that far-reaching emissions reduction is necessary. The challenge is to attain it at a reasonable social cost, which requires cost-reducing progress on environmental technologies. Present efforts are costly. A large and increasing share of the countries' Gross National Product is spent on environmental technologies. The expenditures cover 1.8% to 3.5% of the Gross National Product in developed countries, which is a higher figure than, for example, expenditures on social welfare in many countries. And although smaller, this percentage is growing fast in developing countries. In the future, the expenditures could increase even faster in order to achieve more rapid progress in emissions reduction and to address new issues such as the emission of greenhouse gases.

This book's central question is how to act on the demands for good environment in an affordable manner. There is abundant theory about environmental policy and management starting from the seminal works of Baumol and Oates (1975) and Siebert and Antal (1979) down to the recent handbooks written by Pearce and Turner (1990), Tietenberg (1994) and Perman et al. (1999). A key role in answering this question is the development and use of environmental technologies that prevent degradation of environmental qualities. These technologies integrate techniques, products and methods for applications in environmental policy and management. Much economic theorizing about environmental technology can be found. However, the empirical support of this is scarce, drawn mainly from cases found in a few recent valuable volumes, such as Grubler et al. (2002), Weber and Hemmelskamp (2005). This book starts with practice and moves beyond cases to link theory and practice. Authoritative databases, statistics and simulations are used next to companies' and policy cases to explore and test theory and elaborate on tools for policy makers and management. This is the practice in the European Union. Most empirical data come from the Netherlands, which has been at the forefront in the environmental area for the past 50 years. The book aims to help avoid some of the past's costly mistakes regarding the need to curtail emissions in the future.

The book provides an optimistic answer to the challenges of far-reaching emissions reduction. It is argued that many technologies are already available to reduce most emissions by 75% to 95% and some types of emissions by an even higher percentage. Far-reaching emissions reduction is often technically possible, though there are important exceptions that must be addressed. The main problem is that many technologies are costly. This book illustrates that progress in effectiveness and efficiency has been achieved in some areas and argues that more progress can be expected through sound decision making.

Authorities, companies, expert centers and individuals, basically everyone, can develop technology to create a novel product, service or method (invention) that can subsequently be produced for sale to users (innovations) and thereafter disseminated on various markets (diffusion). Innovation or "doing things differently" (after Schumpeter (1935), 1989, page 59) entails the use of technology, in this case, in environmental policy and management. In line with Schumpeter's analysis of innovation that underscores the prominent role of risk-taking entrepreneurs, the book addresses problems in the decision making done by authorities, users of technologies that are polluting firms and innovators who are developers of environmental technologies. The issue is how to make decisions about development and the use of environmental innovations regarding uncertain demands for environmental qualities and uncertain costs and performance of the technologies. These decisions must consider the fact that environmental technologies are expensive to buy and use and rarely contribute directly to the profitability of individual companies. Rather, they are rewarding to society as a whole.

The deliberation of users is mainly technology selection, which means comparison of the options available from the past and the new ones expected to enter the market in order to comply with the demands at lowest cost. The first difficulty in decision making is the size of investments. Environmental technologies are usually capital goods that need large investments and require infrastructure for operations. For example, politicians in local authorities could face investments in wastewater treatment that exceed the municipality's total annual budget, entailing an even larger outlay in sewage, which could require unpopular taxation to cover the costs. Secondly, managerial decisions by firms are challenging because investments that rarely contribute to profit are required for a license to operate and, thirdly, because know-how is needed that usually does not contribute to productivity. For example, a company could be ordered to clean soil at a cost but without profit. In the same way, a retailer could be requested to change its assortment because of the noxious compounds found in the materials of which he had not been aware.

The innovator's decision about development of an environmental technology is even more troublesome. The innovator would need to invest in research, development, demonstration, production and marketing of an innovation that remains unrewarding for many years before the innovation can be introduced to markets and be sold due to the policies demanding additional emissions reduction. The rewards due to sales remain uncertain because performance of the realized technology and future demands are difficult to assess beforehand. However, the innovator has a chance to reap high rewards due to the advantage of being the first in the market, i.e., the first-mover. It is a typical "low probability, high reward" deliberation. Next to this, its know-how is usually so specific that it is difficult to use for other purposes. Furthermore, the innovator faces a race to patent the invention because the relevant know-how needed for the innovation is partly in the public domain and can be used by competitors as well. In addition, the innovator must face the possibility that much of the know–how will remain in the public domain because it cannot, or even may not, be protected by patents and therefore can be copied by others. The consequence is that investments in know–how remain low because private expenditures in technology development are not sufficiently rewarding and the innovator must generate a lot of specific expertise. These "spillover" effects that cause underinvestment in know–how are particularly relevant for environmental technologies because environmental qualities are considered to be not directly relevant for production, largely publicly regulated and know–how is therefore largely generated in public institutions (Jaffe et al., 2005).

The book's starting point is that environmental innovations are realized for profit. Decision making must consider uncertainties about future costs, qualities and sales of innovations, competing alternatives, different user strategies and various policy making instruments. The decision making is presented from different angles because various interests are involved.

We start (Chapter 2) with the presentation of the theories on technology development that aim to explain why environmental innovations emerge and how can we progress towards low–cost environmental technologies. The mainstream, the evolutionary and the behavioral economic theories on technological change are discussed.

Secondly (Chapter 3), we address the policy makers who aim to enforce emissions reduction at the lowest possible costs, which is usually translated into selection of low–cost technologies able to comply with prescribed emissions reduction at sources. The question is raised: how can policy makers assess the cost of implementing new technologies at various emission sources without having experience with them? The challenge and a few cost-assessment methods are discussed.

Thereafter (Chapter 4), we take the perspective of financiers of environmental innovations, which can be authorities and executives who must decide to invest in years of research, development, design, demonstration, production and marketing of a new technology, but who stand to gain uncertain revenues from sales and those who would like to buy and use the innovations but are uncertain about the benefits. The issue is how to assess the benefits of environmental innovations without knowing about the costs of competing technologies at many different emission sources.

Then (Chapter 5) we turn to the cost-saving effects of environmental innovations in the polluting industries to underpin, based on the statistical data, that companies have been able to reduce some emissions down to the levels that do not preclude sustainable development alongside with substantial decrease of the costs. The technological progress towards lower emissions and costs is assessed.

The progress is explained in Chapter 6. It is argued that strategies used by the polluting companies confronted with strict policy demands are crucial at this point. The dilemma is to avoid risks and wait until enforcement of the environmental demands takes effect or to anticipate them through investment in environmental innovation with cost reduction. We discuss the companies' investment strategies that reduce the costs of environmental technologies and can even provide net benefits of the investments.

The possibilities to cope with the consumers' demands in the life cycle of products are presented in Chapter 7. Based on the real-life cases of life-cycle management it is illustrated that substantial cost savings and even profitable options can be reached by a few focused innovative actions in the life cycle of products.

Next (Chapter 8), we look at policy making from the perspective of the innovators. These can be those working in the research and development department of polluting companies as well as the manufacturers, constructors, engineers and consultants who are specialized in environmental technologies. It is shown that the innovators are attracted by the announcement of policy demands that enlarge the market for environmental technologies, but they are risking time-consuming and uncertain policy making.

In Chapter 9, the innovators' view is redrawn regarding negotiations between stakeholders, called "self-regulation". It is argued that the negotiations potentially provide efficient solutions, but that they actually only trigger minor improvements in available technologies because of the innovators' risks. Hence, institutional changes are needed for successful negotiations.

Finally (Chapter 10), we conclude by discussing the possibilities to foster environmental innovations. This chapter reviews the possibilities available to allocate efforts in technology development more effectively and efficiently, as well as the conditions and policies that contribute to technological progress.

2

Technology for Sustainable Development[1]

Debate continues on the question of whether or not it is possible to foster productivity that contributes to economic growth and a rise in income, while at the same time decreasing emissions down to the level that makes it possible to maintain availability of the environmental qualities for future generations. In this chapter, we discuss if it is possible to decouple economic growth and material use (so-called dematerialization) and whether it is possible to steer technologies towards less material use and emissions per unit of produced value (so-called ecoefficiency). A final answer cannot yet be delivered, but the arguments are discussed to pursue more balanced decision making in environmental policies and management.

2.1 Sustainable Development

The debate about dematerialization has been going on for the last few decades. However, the focus has changed. In the 1960s and 1970s, after a few decades of exceptionally high economic growth in Europe, Japan and North America, attention was given to the negative side effects of growth such as fuel and mineral exhaustion, extinction of species, health impacts, and fragmentation of land and so on. It is argued that the materials used in production and consumption disperse as emissions after some time and that the mass-conservation principle (the amount of emissions equals the extracted material and fuel resources) implies dispersion of the materials during and after use. However, the qualities of the dispersed materials change and degrade and the materials cause negative effects on environmental qualities (Kneese et al., 1970; Ayres, 1978). Some environmental economists argue that economic growth and the availability of environmental qualities are

[1] Parts of this chapter were published in Krozer and Nentjes, (2006), An Essay on Innovations for sustainable development, *Environmental Science*, 3(3):163–174. I am grateful to the Taylor and Francis Group for permission to use the material.

fundamentally incompatible. Many others assume that environmental qualities can only be sustained under far-reaching income distribution and radical technological changes in favor of environmental qualities. They point out that the degradation of environmental qualities causes welfare and productivity losses that should be incorporated in economic indicators like prices and income as this provides a more realistic view on welfare development in national accounts (Leipert 1985; Daly and Cobb, 1989; Hueting, 1990).

Contrary to environmental economic views, the mainstream (neoclassical) scholars expect that, in theory, the economic structure does change. More specifically, it is thought that, in theory, the non-reproducible factors (environmental qualities) can be substituted by the reproducible ones (labor and capital). Following this, it is underscored that global welfare can grow without depleting environmental qualities if the changes in economic structure proceed faster than the use of the non-reproducible resources (Kuipers and Nentjes, 1973; Solow 1974, 1977). The process of substitution can go on forever due to unlimited knowledge and can be directed towards the technologies that are based on non-scarce and renewable resources, recycling and the creation of new resources. In theory, the only physical limitation is the influx of solar energy that is so large that it is almost unlimited for human development (Weitzman, 1977; Gregori, 1987). It is stressed that the reservation with respect to the substitution of the non-reproducible by reproducible inputs is not so much availability of natural resources but rather emissions that cause pollution. This is because pollution damages environmental qualities, in particular biological processes like biodiversity, health and so on. In a precautionary mainstream view it is accepted that technological progress can create value at decreasing emissions and it is underlined that far-reaching emissions reduction is the prerequisite for welfare growth. In this view, emissions must be decreased to the level that does not distort availability of environmental qualities. That means a reduction by a factor of ten or more in 20 years in Europe and North America in order to balance the growth of materials in use (Perrings, 1991; Nentjes, 1990, Klaassen and Opschoor, 1991; Weterings and Opschoor 1992).

In the 1980s and 1990s, it was widely acknowledged that maintaining good environmental qualities was a precondition for welfare growth. This view was rooted in the political debate occurring under the term "sustainable development". This term, introduced into international politics by the World Commission on Environment and Development, is defined as: "development that meets the needs of the present generation without compromising the ability of future generations to meet their own needs" (WCED, 1987, p. 43). The stewardship of economic development became the focal point of attention. The idea of stewardship through co-operation between market interests and public decision makers addressed the need of technological change towards low-emission methods, renewable resources, resource saving, durable products, low-input production and consumption, non-toxic means, effective space use, and so on. Stewardship in entrepreneurship is underlined with biological metaphors such as sustainable metabolism, tree-like companies, industrial metabolism, green business and industrial ecology. These ideas are translated into a metaphor known as "triple bottom line" (profit, people, planet) that balances income growth, distribution of wealth and environmental qualities (Reijnders, 1984; Winter, 1987; Ayres, 1989; Elkington and Burke, 1990; Graedel and Allenby, 1995). Economic models have also been developed to emphasize the possibility of sustainable development under the assumption of progress in dematerialization and the rapid increase of ecoefficiency among technologies (Hartog and Maas, 1990; Meadows et al., 1991; Duchin and de Lange, 1994).

The empirical findings gained from 1970 to 1990 do not support overly optimistic views. The positive trend is a growing share of labor-intensive services despite increasing labor costs. In addition, there is a trend towards increasing value added in the manufacturing process in many industrialized countries. This trend causes reduction of energy and material use and emissions per unit of output in several countries including Germany, Japan and Sweden. The trend cannot be explained solely by the export of the most resource-intensive and polluting manufacturing processes from industrialized countries to developing ones. The main factors are changes in the manufacturing towards products with high value added and the use of environmental technologies that reduce material uses and emissions (Jänicke et al., 1986, 1997). Another positive sign of the decoupling between economic growth and the degradation of environmental qualities is a significantly negative correlation between the emissions connected with combustion—CO and NO_x—and the countries' Gross National Products. This trend suggests that there is a global shift towards less energy-intensive economies that is explained by the diminishing share of manufacturing in the global economy (Selden and Song, 1994). The findings on changes towards more services and high-value manufacturing that contribute to dematerialization invoked the view of an autonomous, long-term, global trend from the less-developed economies that are largely based on material-intensive manufacturing towards the more-developed economies that are largely service based. A hypothesis is put forward of the so-called "Green Kuznets-curve" that advocates a relation between income and pressure on environmental qualities. The relation is expressed by an inverse U curve. This suggests low pressure at the low-income level due to little manufacturing found in low-income countries, growing pressure to increase income because of more manufacturing in medium-income countries and less pressure in the highest-income economies because of the presence of more services. The message is that the autonomous trend towards higher income stabilizes and ultimately reduces the pressure on environmental qualities. This view is scrutinized in view of poor data on material use and pollution in most of the world's countries and failure to accommodate international material flows in the statistical accounts (Ayres, 1997; Bruijn and Heintz, 1999). Studies examining international material flows support the criticism. Per capita material use and emission, including the trade with materials, shows an increasing trend in Germany, Japan and the Netherlands, and stabilization in the United States but at a much higher level than in Europe and Japan (Bringezu, 1997; CE 2002). At best, it can be argued that the pressure of emissions on environmental qualities does not increase as fast as economic growth, but that does not mean that the impacts do not increase as fast or even faster, because growing pollution can cause fast deterioration of some qualities, such as biodiversity or atmospheric structure.

The discrepancy in findings about the positive changes in economic structure due to more service and the sluggish decrease of emissions pressure on environmental qualities can be explained by complementation in consumption. Complementation in consumption means that people tend to add various uses to products and services as their income grows instead of substituting one product for another. As a result, product consumption decreases very slowly. There are many examples of complementary uses as a function of income growth, including radio and television, paper and computers, telephones and e-mail, and so on. During the upswing, people spend on various uses that are only apparently substitutes for each other, for example people spend more on architects' services for new houses as well as more on professional maintenance of old houses and they buy more products to maintain houses by themselves, albeit the growth rates on the uses differ. Consumptive uses can rarely substitute for each other, but some uses grow faster than others as a function of income growth. Differences in the rate of growth cause the consumption patterns to change. Mobility is illustrative; mileage data for the last century show that the share of old transportation methods (ship and rail) decreased in favor of the new modes (road and air), but all transportation modes grew, albeit some faster than others (Nakicenovic, 1991). The complementation can be explained by the fact that product use is closely linked to the uses of many services and products in a product–service combination, for example, car use calls for roads (products) and regulators (services), and the use of televisions requires electricity and broadcasting. Therefore, it is difficult to substitute a product or a service because it entails a chain of changes in use. In addition, the positive effects that more services in an economy have on dematerialization are counteracted by gradual materialization of services. This means that the share of labor involved in the use of a service at a given income level is gradually substituted by capital and materials (Uusitalo, 1983; Krozer et al., 1996; Gatersleben and Vlek, 1998).

Dematerialization of the economy due to increased services in economies is an important trend for environmental qualities. It progresses due to changing consumers' preferences almost independent of companies' decisions and policy making. However, the growing share of services alone is not enough to limit the growth of emissions as they expand despite an increasing share of service, let alone a reduction in emissions by a factor of ten to twenty in the next few decades. In addition, an increasing ecoefficiency in production and consumption is needed at a rate that is much higher than production growth to compensate for pollution's negative effects on environmental qualities. The key factor for ecoefficiency in consumption is the fast development and diffusion of environmental technologies over a broad range of activities in production and in many types of consumptive uses. The issue becomes whether it is possible to foster technological development towards low-emission patterns at a much faster rate than the economic growth. This issue is discussed in view of economic theories: the neoclassical, the evolutionary and the behavioral theory.

2.2 Theories on Environmental Technological Progress

The economic theories on technology development provide different answers on the possibilities to foster changes towards environmental technologies. The differences are connected with diverse points of view on technology development: scarcity and prices in the neoclassical theory, policy decisions in the evolutionary theory and companies' organization in the behavioral theory. The neoclassical theory is so preoccupied with prices that other factors, including possibly even the main factors that influence technological development are less elaborated. Other factors are more explicitly considered in the other two theories. The evolutionary theory focuses on the positive external effects of technologies (spin-off). The means of productivity increase in many sectors due to the application of some closely related technologies (called a technology path or a *filiére*), thus suggesting that cleaner technology in a sector can provide spin-off to many other sectors. The behavioral theory addresses the internal organization that decides upon risk avoidance instead of output maximization. This theory suggests that innovation is an anomaly. These views partly exclude and partly supplement each other. After the review of these theories, we will enrich mainstream environmental economics based on the neoclassical economic theory with views from the other two theories. We discuss the theories primarily from the perspective of decision makers contemplating what guidance they offer in the quest for progress in ecoefficiency. The behavioral view has been largely neglected in the literature on environmental innovation and it is extensively covered here because it offers a fruitful approach to understand companies' decision making about environmental innovation, in particular on how incentives affect firms' innovation decisions.

2.3 Neoclassical Theory

The neoclassical theory concludes that welfare losses from pollution are unintended consequences of failures in market organization and in public sector performance. Market failure appears where property rights regarding environmental goods, like the right to use the environment as a sink for pollutants, have been imperfectly defined. Consequently, parties that suffer from pollution cannot develop trade with parties who benefit. Since a market for scarce environmental goods does not emerge spontaneously, scarcity remains under-priced suggesting to polluters that there is no scarcity at all. One option for the victims of pollution is to try to make polluters liable for environmental damage. Such lawsuits clarify the rights of respective parties, laying the basis for negotiations and the elimination of market failure. However, if the sources and the victims of pollution are many, remedial action through private law breaks down. In such circumstances, which are typical in high-income and developing countries, control of pollution is a public good and a task for national government (Angel, 2000). Public-sector failure occurs when governments fail to take appropriate action. In the neoclassical view, the consequence of such failures is that the environmental scarcities are not signaled, either in price of pollution or in any other restrictions on pollution that can be imposed by regulations. If pollution is free then the economic incentive to contain emissions is lacking. In addition, so is the incentive to invest in research and development of environmental technology that would provide the means to reduce or prevent pollution. As a result, the costs of development and procurement of environmental technologies are higher than the benefit of less damage due to lower emissions that could be achieved by the installation and use of the technologies. The argumentation follows that if the market mechanism does not specify the liabilities for the external effects sufficiently, then the policy makers should avoid pollution by emissions reduction. Policy makers can use various instruments that trigger companies to reduce emissions. Basically, policy makers can put a price on emissions or restrict emissions by standards with a maximum emissions allowance. The price and the standards invoke the use of environmental technologies to reduce the polluting activities, because it becomes beneficial, or in the second case to reduce emissions. In its policy advice, neoclassical economics has a clear preference for instruments that mimic a market, which is setting a price on pollution. Placing a price on pollution has two effects on polluters: it signals environmental scarcity and it provides polluters with an incentive to take action, while leaving them flexibility in their search for the best approach, including the search for and the development of new more effective technologies (e.g. Baumol and Oates, 1975; Pearce and Turner, 1990, Tietenberg, 1994). Following the neoclassical view, production for valuable markets as a function of inputs and production to prevent damage as a function of emissions reduction are differentiated (so-called joint production functions), which implies that companies confronted with environmental regulations must deliberate between use of additional environmental technologies and limitation of the most polluting activities (e.g. Duchin and Steenge, 1999).

This exposé of environmental economics is an application of the neoclassical theory on induced technological development. In this theory, perfect substitution between inputs is assumed in the long run. The substitution enables one to maximize output at the lowest input prices, whereas the prices are determined by scarcity of inputs. In this way; companies develop and apply technologies as a result of an exogenous set of input prices as opposed to autonomous technological development driven by the progress of knowledge that is an endogenous factor (Heertje, 1973; Stoneman, 1983; Grilliches, 1996). The theory on induced technological development is supported by many empirical studies among others in agriculture. The studies show that the high prices of inputs (measured by the relative price of agricultural inputs) invoke efforts in research and development (R&D) to develop technologies (measured by patents) that in turn, reduce the use of the most costly input (Ruttan, 1971, 1982). This train of actions is less clear in the case of environmental technologies. The poor responsiveness of environmental technologies with respect to resource and emission prices is explained by various imperfections, dominance of public interventions such as subsidies for natural resources and polluting products that undermine the positive effects of the prices on the use of environmental technologies, low-resource productivity that limits expenditures on R&D and protectionist measures like patents and import restrictions that create barriers to the entry of new technologies. As stated in the theory, the imperfections limit prices' positive effects on use and development of environmental technology. These imperfections are reflected by low elasticity of emissions reduction with respect to prices (Dasgupta and Heal, 1979).

Despite the recognition that price effects on environmental technology are far from perfect, the theory on induced technological change is widely applied in environmental policy. The theory is instrumental in practice because it relates the effects of policy instruments like emission standards or emission charges with the development and use of environmental technologies. The theory suggests a causal relation between scarcity, price and technological change. However, the theory's empirical foundation is weak. Based on the theory, it should be expected that the lower resource prices increase the use of material in products. In reality, the share of material measured by mass and value actually decreased as did resource prices during the past two centuries. In the 19th century, the prices of natural resources were already so low that the cost of resource use was considered hardly relevant for companies' decisions. In 1848, Mill remarked: "But the crude material generally forms so small a portion of the total cost, that any tendency which may exist to a progressive increase in that single item, is much over-balanced by the diminution continually taking place in all the other elements; to which diminution it is impossible at present to assign any limit" (Mill, 1985, p. 64). Ever since, the real prices of resources has steadily decreased albeit with some fluctuations (Rosenberg, 1975b, p. 229–248; Dasgupta and Heal 1979, p. 439–470), as well as the share of materials per unit of product, measured by real value and by weight, which is found on the national and sector level, as well as being illustrated by product cases (Larson, 1986; Herman, et al. 1989; Tilton, 1991; Wright, 1997). Similar trends should be expected for emissions. It seems that the price of natural resources, or emissions is not the only, and maybe not even the main, determinant of the changes towards environmental technologies.

The theory on induced technological change is criticized within the neoclassical theory. Scholars on the history of technological change have argued that the development of a new technology usually takes so many years that it is impossible to foresee the resource prices at the moment of sales of the technology, because the resource prices fluctuate. Hence, the demand factors, such as prices, are mainly relevant for the improvements of the available technologies (adaptations) because adaptations are less time consuming and the results of adaptations are more predictable than the development of new technologies (innovations). Studies on trends in technology development argue that the relation that is suggested in the neoclassical theory should be reversed. It has been put forward that technological development is largely driven by the cumulative increase of know-how. It boosts productivity that reduces the use of the costly inputs. The decreased use of the costly inputs, in turn, causes downward pressure on resource prices. Technological development thus explained by generated know-how is triggered by dramatic events like overexploitation of natural resources, population changes, wars and so on (Lilley, 1980; David, 1975; Rosenberg, 1977, 1982a). Therefore, distinction between autonomous (endogenous) and induced (exogenous) technological change should be made, although it is not clear why endogenous development is strong in the case of environmental qualities. It is rather odd in view of much higher and increasing labor and capital relative prices. An explanation is needed.

2.4 Evolutionary Theory

The observation of changes in economic structures irrespective of the relative input prices invoked another interpretation of technological development. In the argumentation that is usually called evolutionary theory, the technological development is viewed as the search and selection processes within socio-cultural and administrative frameworks that are determined by prevailing norms. In the search process, various options are presented to solve a problem, for example, to overcome resource scarcity. The selection process includes choosing the most suitable option for a firm or an institution in its specific situation. The final decision is rarely based on thorough economic calculations like cost benefit assessments. Instead it is based on the decision maker's expectation based on prevailing norms that the selected technology can perform well. In this train of thought, it is expected that input-prices have minor influence on technological change in comparison with generated know-how, socio-cultural conditions, quality of management, policy making and so on (Nelson and Winter, 1982; Dosi and Orsenigo, 1988).

Following this argumentation, it is pointed out that a new technology invokes development and use of other related technologies in various sectors, so-called spin-off. For example, development of computers invokes software that triggers investments in the software industry that in turn invokes development of the Internet and so on. The result is that many linked technologies form a pattern, called a path. The related technologies, becoming path dependent, entail many improvements that raise productivity. A positive effect of path dependency on economic development is an increasing return to scale in many sectors. A negative effect is that once an inefficient technology is established, it is difficult to substitute it with a potentially superior technology because of interrelated, past investments. The established technologies have a character of sink costs; so the technology path becomes a pervasive system. Only large investments in a new technology can break down the superior pattern and establish a new one (Arthur et al., 1987; Arthur, 1989; Arthur, 1990). This causes regional and structural repercussions, because many companies depend on each other in clusters. Even partial replacement is difficult because the activities linked with the established technology do not fit with the new technology and consequently the whole system of businesses collapses. Forceful policy interventions are needed to introduce a new pattern or cluster, like subsidies for co-operation between companies, new stringent regulations and public investment (Malecki, 1991). An illustration of the problem is the idea of substituting a hydrogen-based energy system for the present fossil-fuel-based energy system that would require huge capital investments in a new energy infrastructure to replace the present infrastructure that would have to be dismantled. Technology is locked in because the huge capital investments made in the past for the infrastructure and organizations crafted on the established technology are sunk costs.

The evolutionary theory presents a convincing image of technological development, but it does not explain the relations between socio-cultural factors and technological change, nor does it provide arguments on how to assess the changes in technological patterns beforehand. Some call it "theorizing" because causal relations are absent (Nelson, 1995). The pros and cons of the neoclassical and evolutionary theories are not discussed further as they are lively debated in literature (Stoneman and Diederen, 1994; Metcalfe, 1994; Balman et al., 1996; Ruttan, 1997; Dosi, 1997). We focus on the application of the evolutionary theory in environmental technology.

Advocates of evolutionary theory on environmental technologies make a distinction between two development patterns. The cornerstone is the dichotomy between treatment technologies and process-integrated technologies that make the pattern with the technologies for emissions treatment that are called add-on or end-of-pipe technologies and the pattern of technologies for emissions prevention that are called cleaner technologies. In decision making, emissions treatment is called a curative approach as opposed to the preventive approach offered by cleaner technologies. The difficulties involved in reducing emissions are primarily explained by the dominance of technological patterns from the past that comprise emissions treatment. It is assumed that the technologies have been selected without adequate, overall consideration of environmental issues and that the substitution of the treatment technologies by the process-integrated technologies is imperfect. These scholars even suggest that environmental policy can be a major cause of the imperfections because it enforces far-reaching pollution reduction in the short-term that promotes emissions treatment instead of longer-term policies that can invoke cleaner technologies. Along these lines, various socio-cultural factors are presented to explain why companies favor treatment technologies instead of process-integrated technologies. These factors include: poor awareness and information about environmental qualities, imperfect selection of solutions by management as well as policy focus on standards in licenses that prescribe treatment instead of process integration (Quakernaat et al., 1987; Mensink, et al., 1988; Schot, 1988; Christensen, 1991; Saviotti, 2005).

Forceful policy interventions are argued to introduce environmental innovations. The State of California's regulation in the US on electric cars, so-called zero-emission vehicles regulation, is used to illustrate the fact that policies can trigger clean-technology development. It is argued that the regulation can force the environmental innovations that, in turn, provide competitive advantage to US car manufacturers. This occurs because they gain experience in the state on how to sell electric cars in other parts of the world (Kemp, 1995, p. 262–282). However, this example also illustrates that predictions are risky and fail. The case shows how non-compliance with the regulations has invoked several successive policy amendments that weaken its stringency and that discouraged the innovative spur and delayed implementation. Nevertheless, it is suggested that time and again decisive policy makers and managers can change the pattern in environmental technologies from the curative approach (emissions treatment) to the preventive approach (using integrated technologies). Breakthrough innovations are advocated that would provide positive effects on the economy and the environment (Weizsäcker, 1998, Weaver et al., 2000). Various policy interventions in R&D, intermediary organizations and implementation of environmental policies are encouraged to reach breakthrough innovations, such as: assistance with initiatives, funding of the development of cleaner technologies and strict regulations (Arentsen et al., 2001; Kemp and Moors, 2003).

Evolutionary theory as applied to environmental technologies argues that there are choices in patterns of technological development and the metaphor of cleaner technology paths as a result of decision making is appealing. The theory is attractive because it suggests managerial capability to steer technologies towards sustainable development, but it does not outline which mechanisms cause shifts towards cleaner technologies and why present decision makers can make better choices than decision makers in the past. Moreover, the dichotomy between "good" and "bad" innovations is dubious because it cannot be predicted whether a "clean" technology becomes "dirty" during use and vice versa. It is uncertain beforehand if the potentially cleaner technology actually contributes to emissions reduction or vice versa, if a dirty technology can become a cleaner one due to additional actions. Experience shows many unfortunate decisions and changes in mindset in decision making. For example, trains were the example of "dirty" technology in the 1920s. However, nowadays they are considered to be the clean transport system. Nuclear power plants that were expected to provide unlimited, clean energy in the 1950s were found unacceptable in the 1980s. And recently, return packaging has been assumed to be environmentally sound, but assessments and experiences in many countries undermine this perception. The advocacy of the breakthrough towards clean patterns is also disputable because dissemination of incremental changes can be very effective. For example, a 10% annual rate of technological progress towards more effective technologies reduces emissions by a factor of five in 15 years and by a factor of 10 to 20 in 20 to 25 years. The rate is realistic regarding the experiences with technological development that indicate progress in effectiveness during many decades, like in shipping (Rosenberg 1982a). Such a high rate of the effect-increasing technological progress without breakthrough technology is also found in environmental technology. For instance, the tenfold higher energy performance of locomotives in the 20th century measured by the pulling power per unit of mileage (Heel and Jansen, 1999), or in the air fleet that on average reduced fuel use per passenger kilometer by half during the period 1970 to 1990 (Flemming, 1996). Moreover, treatment technologies became very effective. In the last few decades, treatment technologies have reduced many emissions by a factor of 4 to 10 at similar or even lower costs through an innovation with subsequent adaptations such as the almost 90% reduction of biological matter due to improvements at wastewater treatment plants and more than 95% SO_2 emission reduction through better ventgas treatment.

Far-reaching emissions reduction can be achieved through innovations and adaptations of available technologies, both by treatment technologies and by process-integrated technologies. These patterns can be seen as competitive technologies at the moment of decision making, but all of them can be equally good. If policy makers, following the evolutionary view, decide to subsidize heavily process-integrated technology because they need a breakthrough with an uncertain effect of the subsidy on environmental performance this then contradicts the rule that rivaling technologies should compete on a level playing field. As a result, breakthrough technology may emerge as the competition's winner, even though it is not actually the most environmentally benign solution. At the same time, adaptations of the available treatment technologies could be triggered by regulations with hardly any subsidies. The need for environmental innovation can be motivated by high costs and poor effectiveness of environmental technologies available at the moment of decision making. However, decision makers must also consider that it is uncertain whether a new technology becomes superior after installation and if it performs well during use (Rothwell, 1992). Doubts about the necessity of breakthrough from treatment technology to the process-integrated technology path should not be considered a plea against environmental innovations, but rather as a way to foster innovations in areas that lack effective and efficient environmental technologies.

2.5 Behavioral Theory

The behavioral theory provides a convincing presentation of companies' decision making on the development of new technologies by addressing firms' organization. In the exposition of the theory, the work of Cyert and March (1968) on the organisation of firms is linked with decision making on environmental technology. In their seminal work *A behavioural theory of the firm* Cyert and March from 1963 criticise neoclassical economics for modeling the firm as a single-minded profit maximizer, possessing all relevant information on the options from which it can freely choose, and acting without internal co-ordination problems, as if it were one person. In contrast with the holistic conception of the firm of neoclassical economics, the behavioral theory has a pluralistic view. It sees the firm with its different functions at different levels as a conglomerate of interest groups, each with its own specific objectives imperfectly coordinated by the firm's top management, because of incomplete information and control. We shall clarify this view, using three key concepts: satisfying behavior, organizational slack and conflict resolution.

Satisfying behavior means that a department's objectives are set as aspiration levels, mainly determined by extrapolation of past achieved results. It can be, for example, that the target for marketing department's sales volume is increased once sales targets of the last period have been met. If an objective is not achieved, options to solve the problem are considered one by one starting with the least incisive option and within the department where the problem emerged. For example, a drop in sales has to be solved in the first instance by the marketing department. The search for options stops if an option that promises to meet the aspiration has been found. When circumstances become even more difficult and the search for alternative options has to be widened, solutions requiring more drastic changes and involving higher risks are taken into consideration. Conflicts between the potentially competing objectives of different departments in a firm, for example, aspired sales level and profit aspiration, can be avoided because each group does not go for the unknown best but rather for a satisfactory outcome, given its aspiration level, and because organizational slack, also known under the name X-inefficiency, offers a buffer. Slack is expenditure that is not really necessary. It is a form of waste that tends to rise in good times when a firm's management units achieve their aspiration targets. The search for better options, which starts when aspiration levels are not achieved, will result in detection and reduction of X-inefficiency or absorption of slack as Cyert and March (1968) call it. A third element that helps to avoid outright confrontation is the established internal procedure for decision making. In particular, this is the guideline that if an aspiration level is not achieved the unit whose objective is not achieved must come up with a solution. For example, a drop in sales below the aspiration level can be countered by the marketing department through slack reduction. This means a more effective marketing effort with an unchanged budget. Other groups come in when the problem cannot be fixed locally, for instance, production is adjusted when stocks increase due to lower sales. The company's top management acts when problems at lower decision levels accumulate and its major objective—the aspired profit level—is not achieved. The selected solution, if adequate for the objective, is then internalized in management procedures such as internal quality assurance. The solution becomes the preference routine in management of the company.

From the behavioral theory emerges a picture of the firm as a plural organization that relies on conventional solutions and is sluggish in adjusting to changing external circumstances. Actions do not automatically lead to the optimal solutions predicted by neoclassical economics. In short, firms decide under bounded rationality.

March (1989) has investigated the implications of the behavioral theory for innovation. In line with his notion of the firm's behavior as satisfying aspiration levels preferably by conventional actions and considering a restricted number of options for solutions in a hierarchical order, he argues that innovations are not the result of spontaneous, voluntary actions. On the contrary, the established positions of management units in decision making form large barriers to change because innovations are perceived as a risk to the unit. The conventional mechanisms for problem solving dominate, thus avoiding the risks involved in exploring new ways. Innovations come into sight when several possible solutions are considered and are expected to fail. In his view, innovations need organizational changes that enable pop-up, new views such as specialized development units to generate new ideas and invoke changes. He calls for the functions that develop a playful process in decision making to invoke innovations ("foolishness"). Studies into innovation processes confirm that risk taking is rarely spontaneous or directed by management or by calculations. Instead it is a necessity in view of resource scarcity, strict regulations or tough competition. Innovations are rarely deduced from technological patterns in the past and the innovators often act apparently irrationally, driven by self-fulfillment ("with guts"). This view suggests that companies' decisions to innovate are driven by entrepreneurial initiatives that cannot be derived from the past because they distort the existing patterns. Hence, an innovation is uncertain and rarely predictable by policy makers, though clear aims and the expectation of rigorous enforcement are favorable conditions for innovation. Thereafter, numerous adaptations during decades and even centuries follow (Rosenberg and Birdzell, 1986; Coombs et al., 1987; Allen, 1988).

The argument of the neoclassical and the evolutionary theory against the view of trial and error in innovation processes presented in behavioral theory can be that more preparatory work reduces the uncertainties, although some expenditures are needed to generate information and to negotiate with the relevant internal and external interests (transaction costs). Hence, it is argued that more expenditure on transactions could provide an even larger benefit, because better solutions would be found. However, this view is challenged by studies on decision making processes. The studies show that decisions on controversial issues are often delayed to keep track of traditional patterns from the past. Higher expenditure on the deliberation about alternatives often does not help, because more deliberation increases companies' transaction costs, but cause an unanticipated effect: the demand for even more deliberation. This means that uncertainties encountered in decisions about innovations can only be reduced by more information or negotiations with various interests but they cannot be prevented (Colinsk, 1996). The acts of consultancy, brokerage and accountancy have only a limited contribution to risk reduction in decision making. Trial and error is unavoidable.

Building on earlier work (Klink et al., 1991), a view on innovation can be brought together and integrated into the framework of behavioral theory on the firm. Under the conventional policy, direct regulation with mandatory performance standards that can be met with technologies available from the past, the task of environmental management is to comply with the standard at the lowest cost. Environmental management is a technical routine task as the requirement and the technologies are largely preselected by policy makers. When it comes to acquiring the environmental license allowing operations to start, the task of installing the available technology does not affect the firm's profitability because the technologies are preselected that are expected to be affordable for the firm. Hence, the firm's decision is delegated to an environmental unit low in the management hierarchy, possibly a sub unit of the production department. The task can be left to the environmental unit. The firm's top management remains at a distance; it has no motive to interfere. It is not a climate favorable to environmental innovation because the environmental unit has no authority to make R&D decisions. In addition, if it considers proposals to do so, the units higher in the company's hierarchy will not be able to explain how that unconventional option, not belonging to the firm's core business, contributes to the firm's prime objectives. Once the license is acquired and the environmental technology operates, the solution is incorporated as a routine and management attention diminishes because the problem is assumed to be solved.

The position of environmental management would become different, however, if a drastic change in policy were to bring new and very strict regulation, demanding such high investment in the preselected available technology that the costs threaten to depress firm profits below top management's aspiration level. Even more threatening is the exceptional step of technology-forcing performance standards that must be achieved within a number of years with sanctions for non-compliance. A problem also occurs when competitors are expected to accrue a share of the market by provision of the products that comply with policy targets or satisfy customers' growing demands. The environmental unit is unable to fix the problem on its own and other departments must be involved. This is a task for top management to initiate and coordinate a search for options to address the problems. Among the first, least-costly options are lobbying for less-stringent regulations. Another can be postponement of the compliance date, criticizing competitors and taking legal action to delay compliance. When these options are not expected to deliver the aspired recovery of profits, a range of more incisive options has to be considered. These can include starting R&D on new, more effective environmental technology or participation in a joint innovation project. R&D necessarily involves the production department because new environmental technology may require adjustment of production methods. The procurement and sales departments have to participate if the input mix, in particular, raw materials and fuels must be changed and product characteristics and possibly even product image have to be modified. The more that different departments become stakeholders the more necessary it becomes to move the problem up the firm's hierarchy and to face and accept the risks related to unconventional solutions. Innovation is an option that is not chosen spontaneously because it is costly and risky. Only under circumstances of urgency and potential high rewards from a successful R&D investment can one expect a firm to accept the costs and risks inherent to such a strategy and innovate. The more the option means having to reduce or shift production as the way to meet environmental demands, the greater the probability that environmental innovation will be attempted. Depending on the perceived option, R&D may focus either on ambitious improvements in add-on technology, a total redesign of production, or an intermediary solution. In short, behavioral theory of the firm suggests that environmental innovations can only be expected if high-ranking management senses the urgency and anticipates tough environmental demands in the future with promising market conditions.

2.6 A Framework on Environmental Innovations

In light of these theories, the question is how to direct technology towards higher income at lower emissions levels in order to increase greatly the present ecoefficiency. All the theories suggest such a possibility. Analyses diverge with respect to the mechanisms that invoke technological changes and therefore policy recommendations differ. Neoclassical theory argues that technologies can flexibly be adapted to the changing input prices by input substitution. Neoclassical economists emphatically advocate a high price on emissions to reflect scarcity of good environmental qualities. This would invoke development of effective environmental technologies and reach the lowest-cost solutions through competition between the technologies. Evolutionary theory views technological development as a process of lock-in by the patterns created in the past. To force a break through of the existing patterns, policy makers have to choose for new, cleaner options and support technology development in that direction by using a full range of policy instruments such as subsidies, technology-forcing enforcement and so on. Behavioral theory describes technology as an entrepreneurial instrument embedded in the company's organization that essentially resists changes unless the available solutions are insufficient to attain the organization's objectives. Forceful, external demands for environmental qualities in combination with well-informed and creative internal management are necessary requisites to innovate. At first glance, the views seem to point in different directions for decision making. However, closer examination suggests that the neoclassical theory provides a good starting point for a theoretical framework on environmental innovation. The postulate that a price on pollution affects the perception of scarcity is not disputed but the causality that the price determines the development of environmental technology is arguable.

Neoclassical theory's argument that input substitution in the long run is supposed to be perfect is particularly unconvincing regarding the finding that material-intensity in economy decreases even under decreasing resource prices. The finding can be explained by the engineering theory on loss prevention. The theory explains inefficiencies in production by the occurrence of losses in production. A production is described by the output function of heterogeneous inputs to perform a demanded output (a product). On the assumptions that the goal of production is a qualified output goal and operations in the production are stochastic events created by human skills, the outputs are described by factorial inputs (Leeuwen, 1988). It is formally:

$$N_o = (N_i! + 1) \tag{2.1}$$

with N_o outputs and N_i inputs.

The production function of three inputs ($N_i = 3$) can provide seven outputs ($N_o = 3! + 1 = 3*2*1 + 1 = 7$). Out of seven outputs only one is the qualified output (product), whereas all the others are side products that can be waste. For product maximization, six out of the seven outputs must be prevented. The production with four inputs can provide 25 outputs with only one product and 24 outputs must be prevented, and so on. The theory postulates that the number of inputs determines, *ceteris paribus*, the probability of loss and efforts to prevent it. This illustrates how complex any manufacturing process is because even a limited number of inputs provide so many outputs to prevent that 100% valuable output is only attainable through endless experimentation and operational adaptations in production. Scholars on technology change confirm the complexity of manufacturing entailing imperfections in operations that invoke endless engineering experiments to improve complex capital goods such as ships and planes (Rosenberg, 1982b) as well as daily-life utensils such as knives and forks (Petrovsky, 1994). The necessity of experimenting explains autonomous technological change towards lower use of materials by the necessity of loss prevention. Technological change can go forever, irrespective of prices, because any change of inputs in the production brings an exponential increase of outputs that must be prevented.

The engineering theory can be related to the neoclassical economic theory. Input substitution is essentially an innovation. It entails costly adaptations to prevent losses. It implies that those inputs are substituted that pose the risk of the most costly adaptations. The inputs that are difficult to adapt are not necessarily the highest price inputs as neoclassical theory suggests. If input price is high and an even higher price is expected in the future, for example because of emerging scarcity, it can be worthwhile to change the input despite the high additional efforts. The input substitutions or innovations need subsequent adaptations that go on during decades or even centuries to reduce the losses. However, if the price is moderate or no price increase is expected then the costs of input substitution can be perceived as too high. Adaptations of the available technologies follow to reduce wasteful outputs. This process of innovation entailing many adaptations explains endogenous technological progress in environmental affairs that is underlined by the evolutionary theory. This implies that technologies become less wasteful per unit of product as they mature.

The decision making on environmental innovations faces uncertainties. The neoclassical theory and the behavioral theory applied to environmental innovations underline the importance of the demand for environmental qualities that maintain upward pressure on emission prices. The demands, however, are also uncertain because ownership of environmental qualities is imperfectly distributed. Even under the heroic assumption of lasting upward pressure on emission prices there are persisting uncertainties in the decision making. Innovators and users of innovations are confronted with uncertainties. The innovators that supply new technologies are uncertain about the result of effort in technology development and about the sales because the innovations can be wasteful for the users. The users of new technologies are uncertain about the costs and benefits of environmental innovations during the life cycle in comparison to the adaptations of available technology with less uncertainty due to experiences in use. Environmental innovations can be expected in cases where the innovators expect high profits after realization of the innovations and the users expect lower costs or a benefit regarding environmental demands and risks during the life cycle of product in comparison with the available technologies. The sum of the expected innovators' profitability plus the expected savings or revenues, usually called innovation rent, must cover all user costs and risks of environmental innovation. All three factors: companies' expectations about future demand for environmental qualities, the risk perceptions of potential users of innovations and the credibility of innovators, influence progress in the ecoefficiency. The path dependency on the treatment (end-of-pipe) technologies that is supposed to be inferior to the integrated (cleaner) technologies in the argumentation of the evolutionary theory can be explained by high risks changing inputs.

2.7 Conclusion

The presented theories on technology development do not exclude but rather supplement each other. Price signals (neoclassical theory) are important because they co-ordinate decisions, but the effect of prices on environmental innovations depends on policies. If a price is put on emissions, for example, by an emissions charge and effective technologies are already available, then fast dissemination of the effective available technologies should be expected, although it might have little impact on the development of new ones. Subsidies, networks and infrastructure (evolutionary theory) contribute to innovations if the results of interventions are predictable, that is, in cases where there are clear objectives and a few interests that determine progress. Above all, management must sense urgency to innovate despite high costs and uncertainties connected with technology development (behavioral theory). The sense of urgency can be invoked by environmental authorities, liabilities for damage imposed by social groups and exceptionally favorable sales opportunities of environmentally benign products. Policy can create sound conditions for innovation by translating the sustainability perspective into the mid-term objectives that must be attained, but with freedom to act and support technology development.

3

Projecting the Cost of Emissions Reduction[2]

Policy makers deal with many different emission sources that must apply various environmental technologies to reduce emissions to levels that do not preclude availability of environmental qualities for future generations. The question is how can they assess the costs of technologies developed in the past and the new ones at various emission sources? We discuss the possibility of assessing the costs per unit emissions reduction as a function of emissions-reduction percentage in an inventory of emission sources. The costs per unit emissions reduction are called unit costs; similar names are marginal costs or cost effectiveness.

3.1 Best Available Technologies

The costs of policy making cover the preparation and enforcement of environmental policy and the costs of implementing environmental technologies in companies. These costs are substantial measured by their volume and share in the Gross National Product (GDP). They increase quickly as production grows and, in addition, more emissions reduction is demanded. For example, in the Netherlands, the total annual costs of public and private expenditures on environmental policy approached € 4 billion in 1980. This covered about 1.2% of the GDP that year. From 1980 on the costs' average annual growth reached 7.3%, whereas the average GDP growth has been around half of that. As a result, the expenditures cover presently around 3% of the GDP. Expenditures keep growing because small and medium-sized enterprises must be addressed in addition to the large ones already

[2] Parts of this chapter were published in Yoram Krozer, (2006), Projecting costs of emission reduction, in Meijer et al. eds. *Heterodox views on economics and the economy of the global society*, Wageningen Academic Publishers, Wageningen, p.255–268, I am grateful to Wageningen Academic Publishers for permission to use the material.

addressed in the past. Finding ways to curtail the cost increase is a major challenge in policy making.

The challenge is not only voluntary. It is an obligation formulated in the 1999 IPPC Directive on the European Union's environmental policies. The policy is based on the "ALARA principle". This means as low as reasonably achievable emissions reduction. It prescribes that demanded emissions reduction by policy makers should not lead to extraordinary high implementation costs. Hence, the costs and effects of environmental technologies needed to attain the demanded emissions reduction should be well known before authorities enforce emissions reduction at the relevant emission sources in companies. The consequence of the present policy by permit at every major emission source is that the costs and effects of technologies at specific sources should be assessed during policy preparation. During the preparation, policy makers seek expert opinions about suitable technologies and set up projects to demonstrate performance of new environmental technologies that are supposed to attain the policy demands. The demonstrations are executed at a number of emission sources that are expected by the policy makers to be representative of many other emission sources in an industry branch. In practice, the demonstrations of technologies are usually in private companies that are willing to co-operate and in public utilities, like wastewater treatment plants. Another information source are consultants and suppliers of environmental technologies. These, in turn, gain the information by offering services to the polluting companies, usually the large ones, which cover selection and installation of the technologies. Hence, there is information about the implementation of environmental technologies at a few large sources. This information is assumed to be applicable for extrapolation to other sources. Thereafter, the results of the demonstrations and the expert opinions are compiled into inventories of viable technologies. The inventories are used by policy makers, companies and experts to refine sector-wide best available technologies (BAT). The BAT are the environmental technologies that are expected to attain the demanded emissions reduction in an inventory of emission sources, such as an industry branch below the average unit cost of the selected technologies in that inventory. The BAT are compiled in policy documents (so-called BREF) that are used to define emissions standards for the inventory of sources (Sørup, 2000).

Based on the BAT performance, defined by costs and emissions-reduction percentage, policy makers define emissions standards that prescribe the maximum concentrations of emissions at every type of emission source. The emissions standards provide a legal basis to enforce BAT at a relevant emission source in a permit. The companies must implement BAT or their own technologies as long as they comply with the demanded emissions reduction. The implementation provides license to produce. Some countries such as the Netherlands provide a possibility to negotiate a waiver in the license for technology development if a company faces very high costs. A waiver can also be given if the authority and company wish to test a new technology or there is a specific reason to postpone the enforcement of the demanded emissions reduction. This manner of policy making is found in the EU and in many other countries because it suggests assurance of the demanded emissions reduction. The effectiveness of this kind of policy making is not discussed further as it is analyzed in many basic courses on the economics of social problems, like environmental qualities (e.g. Le Grand and Robinson, 1976; Siebert and Antal, 1979; Pearce and Turner, 1990). Inter alia, BAT assessment would also be needed to set tariffs on emission charges below the average unit cost in the inventory in line with the ALARA. However, environmental policies rarely use emission charges, which will be discussed in Chapter 8.

The question is how can policy makers aiming at additional emissions reduction assess the costs of environmental technologies at various emission sources during policy preparation? That is, before they have experience with implementation. More specifically, how can they find representative emissions sources for reliable extrapolation of the demonstrated findings to many other sources? The clue is to determine a method for cost assessments that considers X sizes of emission sources, Y types of sources and Z types of environmental technologies at the sources. An answer is needed to avoid unreasonably high costs at some sources in line with the ALARA principle.

The present policy preparation is biased. Firstly, the expertise is doubtful with respect to additional emissions reduction because the costs and effects of an extra technology cannot be derived directly from experiences with a past technology. Secondly, the expertise is largely based on implementation at existing sources (retrofit investment), because there is little experience with the implementation at new facilities (Greenfield investment). Thirdly, the experts that define the BAT can get information about use of the available technologies from the past not about use of the new technologies. Fourthly, demonstrations of new technologies are executed at companies willing to co-operate but these companies are not necessarily representative of the many other companies in the inventory. Finally, there is limited information about the costs of technologies during use because this kind of information is in the polluting companies' domain. They have no interest in sharing their know-how, but may want to exaggerate the costs because it could prevent stricter emissions standards.

As a result, the reliability of cost assessment in policy preparation is disputable. The current method can indicate the total cost of an industry with many sources because the aggregate of many sources can outweigh the deviations between the sources. However, it does not say much about the costs of any specific company, let alone about the cost at any specific emission source in a company. For reliable cost assessments, it has been proposed to subsidize polluting companies that are willing to share cost information: a higher subsidy as a function of higher unit costs (Lafont and Tirole, 1994). This does not solve the problem. Firstly, reliable information about the costs in use can only be available some years after implementation. Secondly, subsidized companies could seek rents of subsidies instead of low-cost implementation and therefore they could exaggerate the costs. Thirdly, the subsidies could be instrumental for policy making that enforces emissions standards in sectors with a limited number of large sources such as electric power plants. However, for just such sectors there is information from engineering and consulting firms, whereas subsidizing information in sectors with many smaller sources would call for extremely expensive policy preparation. For example, the assessment of technological alternatives solely for EU printing companies (at an average cost of € 10,000 per company with more than 100,000 companies) would necessitate € 1.0 billion in resources. Demonstration projects would be even more expensive.

The answer is to find a method to extrapolate the results found at a few emission sources to many other sources. The cost assessment during policy preparation can be reliable if there is a theory on how to relate the unit costs to an emissions-reduction percentage. The theory would make it possible to define cost functions of emissions reduction by extrapolation of empirical data about a few emission sources to other sources. A usual theoretical proposition is that the scale of emissions determines the costs. The scale proposition leads to an assumption that demonstration of a technology at a few large and a few small sources can be reliably extrapolated to all other sources in an inventory, such as an industry branch. An alternative point of view is that the emissions are so source specific that any specific process variables determine the unit costs. Both views can be valid. The views are tested with empirical data on 28 inventories of emissions reduction. In each inventory, one finds 4 to 256 possible applications of an environmental technology at an emission source. A possible application of an environmental technology at an emission source is called a source technology combination.

Firstly, the theories on cost projection are discussed in this chapter. Then, the escalation of costs is tested in the inventories with one type of emission source and one type of environmental technology, i.e. in the technologically homogenous inventories. Thereafter, a method of cost projection based on process variables is defined on the assumption that policy makers have information about two source–technology combinations in the inventory. Finally, we provide conclusions.

3.2 Theories on the Cost Function of Emissions Reduction

Economists usually embark on the neoclassical theory to define the cost functions. Following this theory, the cost function is derived from an externally given production function at the specified prices of inputs for capital, labor, materials and so on. It is assumed that the production function is well behaved in the long run, which means that there are many inputs that are perfectly substitutable and divisible and the inputs are fully used, that is, production is at maximum capacity. So the lowest-cost combination of inputs can be found at any level of the output. This is presented by a continuous marginal cost function of output. Empirical economic studies on industrial development suggest that technologies are often not perfectly divisible but that the costs increase faster at the lower scale of inputs. This causes a decreasing marginal productivity of an additional input (Rosegger, 1980). The scale assumption is widely accepted in economic theory.

Similarly, emissions are the inputs and emissions reduction is the output of environmental technologies. The production function of emissions reduction covers many substitutable combinations of environmental technologies at emission sources. Following neoclassical theory, one would expect a cost function of emissions reduction that is derived from the lowest-cost combinations of technology at the sources. For example, the cost function of a company with a few emission sources and perfectly divisible technologies in the long run can be: 1 cost unit reduces 50% of the initial emission; the next 0.5 cost unit reduces the next

50%, which means, in total, 1.5 cost units reduce 75% of the emissions. If the scale effects occur, then the unit costs decrease faster than the scale of emissions reduction. For example, 1 cost unit reduces 50% of the initial emissions, 0.8 cost unit reduces the next 50%, with the total 1.8 cost units reducing 75% of the emissions. Such a cost escalation is widely used by consultants to assess the costs of environmental technology at many emission sources in an inventory. The cost escalation is based on the data about a few emission sources of different scale, on the assumption that the inventory is homogeneous, i.e. one type of technology at one type of source. For the escalation, the scale factors are calculated to relate the unit cost with the scale of emissions reduction. This is done in the following manner:

$$\alpha = \ln \frac{\dfrac{cost_2}{cost_1}}{\dfrac{size_2}{size_1}} \tag{3.1}$$

Where $cost_2$ and $size_2$ are the total costs and scale of emission at the small source, $cost_1$ and $size_1$ are the total costs and scale of emission at the large source.

The calculations with empirical data on environmental technologies suggest that the unit costs increase exponentially (logarithmic scale) with respect to the scale of emissions reduction by a factor of 0.8 to 0.2. The small-scale factor means that the costs are sensitive to the scale and vice versa (Remer et al., 1994). If the theory holds, then it provides an elegant method to define the cost functions of emissions reduction because the unit costs can be extrapolated from the demonstration projects at one small and one large source. The extrapolation of the unit costs should hold in every inventory with one type of emission sources and one type of technology that means in all homogeneous inventories of source technology combinations. One can expect that the scale of emission sources determine the unit costs for the homogeneous inventories of source technology combinations. In this theory, the unit costs increase exponentially as a function of scale, but the theory must still be valid.

Cost-engineering studies on production functions provide a different view. It is argued that the technological specifications of the production, the specific process variables, determine the output in mass and cost terms. This means that the cost function describing, for example, the food process is different from one involving the chemical process, almost irrespective of the input prices and scale. The studies suggest that it is difficult to change the composition of inputs with respect to prices because technology fixes a specific balance of inputs to provide a valuable output (Lassman, 1958; Gold, 1975). Porter (1996), for example, identifies more than a dozen variables that influence the balance of inputs in a production chain. Attempts are made to define technology properties that strongly influence outputs. This means defining cost-engineering functions. One attempt was to classify the process variables for emissions reduction by distinguishing technically similar operation units, like grinding, mixing, distillation and so on (Quakernaat and Don, 1988). In this view, for example, the process of *grinding* things like food or wood in various industries causes a similar emission that is dust, which in turn needs one type of environmental technology, such as a *fabric filter*. Other attempts were simulation models that relate the main variables in production with the costs of emissions reduction. This has been done, for example, for the metal-plating industries (Ros and Van der Plaat, 1986) and printing industries (Kuil and Krozer, 1996). These attempts remained untested hypotheses mainly because of difficulties with verification of the simulation results. The verification would require tests in practice, entailing company efforts, but companies are usually not interested in co-operating on stricter emissions standards because of the expected higher costs.

Cost escalation enables one to forecast the costs at many sources with the scale factors and to focus enforcement on the largest sources because the largest sources would reduce emissions at the lowest unit costs, but the cost-engineering functions could be closer to reality. It is intuitively convincing that process-specific variables determine the cost functions. If the example of sulfur dioxide reduction (SO_2) by fluegas desulfurization (one type of technology) at coal-heated power plants (one type of fuel and source) producing electricity (one type of product) is considered with one larger-scale fireplace (one scale) then even in such a homogeneous inventory, the costs depend on the capacity utilization (maximum or less), fuel quality (such as percentage sulfur in coal), distortion of operations by ingredients in fuels, like tar and operational conditions that influence the volume of vent gas (pipe diameter and so on). There is, however, no convincing theory on how to construct cost engineering functions of emission reduction.

3.3 Construction of the Cost Functions

The possibilities to define cost functions of emissions reduction are tested with regard to the scale of emissions reduction and the process variables. The tests are based on the assumption that data about only two source technology combinations are available. The tests are done with the cost-engineering functions of emissions reduction that are constructed with empirical data compiled by H. Heijnes, J. Jantzen, C. Sadee and Y. Krozer from the Institute for Applied Environmental Economics (TME). The cost-engineering functions cover 28 inventories of source technology combinations. The inventories cover many environmental technologies that are enumerated in the Appendix A to this chapter. The models have been developed for policy preparation in the Netherlands, the EU and for the World Bank. The data are from the 1980s and 1990s. The data in models are regularly used in policy making because they are considered the best estimates of the costs and effects of environmental technologies. The data are sufficient to test the possibilities of forecasting, but the data qualities vary because they are a mix of engineering calculations and measurements.

There is data about nineteen sector-wide inventories of source–technology combinations. In the sector-wide inventories, a branch is considered a single source. Several databases and models have been used. The *Basis and Criteria Documents* of the National Institute for Public Health and Environment (RIVM) for Benzene, Cadmium, Phenol, Fluoride, Fine dust, Copper, Polycyclic Aromatic Hydrocarbons (PAH), Propylene oxide, Styrene, Toluene, and Zinc (Apeldoorn, 1986; Koten-Vermeulen et al., 1986; Meulen, 1987; Slooff, 1987; Slooff et al.,

1987; Slooff and Blokzijl, 1987; Ros and Slooff, 1988; Slooff et al., 1989; Cleven et al., 1993). These documents should be considered precursors of the BREF documents on BAT in the EU that have a similar structure. *ICARUS-model* of Ecofys on carbon dioxide (CO_2) reduction (Blok, 1990). *Model on Sustainable Environmental Strategies (Moses)* of the TME on Nitrogen Oxides (NO_x) SO_2, Volatile Organic Compounds (VOC), Ammonia (NH_3), Heavy Metals and Phosphate (Jantzen, 1992).

In addition, nine company-specific inventories were used that cover data about every single emission source in each company. These are obviously much more specific than the sector-wide data. The company-specific inventories were elaborated to support market-based policy on emissions trading with SO_2 and NO_x in the Dutch chemical industry, electric power, basic metal and refineries (Heijnes et al., 1997) and for the Volatile Organic Compounds (VOC) agreements involving the Dutch metal-products industry (Heslinga, 1995).

We describe construction of one cost-engineering function of emissions reduction to reach more understanding of forecasting of the cost functions. The description is based on the sector-wide inventory that is exemplary for the construction of other cost functions during preparation of environmental policies in the EU. In contrast, the level of detail in the company-specific inventories is far beyond EU practice because every single emission source in the company is specified with respect to the technology that can be used, the potential emission reduction and the costs. The construction is illustrated by the cost function of reducing fluoride emissions released into the air in the Netherlands. The example of fluoride emissions (sector-wide data) was chosen because it covers various types of emission sources (industries and households) and several types of environmental technologies (add-on and integrated technologies) with a limited number of source technology combinations. Graph 3.1 presents the cost function of fluoride emissions reduction to air. The unit costs of emissions reduction in Euro per kilogram emissions reduction (€/kg)) are presented vertically. Emissions reduction as a percentage of untreated emissions in the inventory is shown horizontally. The unit costs are a few euros per kg fluoride emissions at the bottom of the cost function up to a maximum of almost € 700 per kg fluoride emission at the top of the cost function. A maximum of 32% of the initial emissions can be reduced. A higher emissions-reduction percentage is technically impossible or there are not enough data about the possibilities. As much as 25% emissions reduction can be achieved by the technologies at unit costs below € 20 per kg fluoride. Up to 26% can be done at the average unit costs of € 45 per kg fluoride. A higher percentage needs technologies that are much more costly.

The graph also includes a streamlined cost function. The streamlined cost function is based on the exponential trend going precisely through the middle of the empirical data. The streamlined cost functions are used to test whether the forecasted cost functions approximate the empirical cost function with the help of only two source technology combinations.

Table 3.1 shows the basic data involved in constructing the cost function of fluoride emissions reduction. The construction is explained in more detail because it is an illustrative example for many other cost–engineering functions, though many cost functions usually end up at a higher emissions-reduction percentage.

The data are largely engineering calculations from the mid-1980s (Sloof, 1989). The data have been collected for policy making on fluoride emissions in the Netherlands. At the time of the study, not all technologies in the inventory had been implemented. Some are only demonstrated or assessed based on information from suppliers. Most data are engineers' estimates. Although it is not possible to validate the accounts, it can be noted that the data were checked by many experts from industry and authorities. Hence, the data should be considered the best estimates available at the time of policy preparation. In a few company-specific inventories, one can find a few extreme values, like extremely high costs of NO_x emissions reduction in basic metal (€ 15 000 per kg) that is related to an irrelevantly small emission source (2 kg). A few extremely high data are excluded from the analysis to avoid erroneous or irrelevant deviations. For the CO_2 only net cost of source technology combinations are used in the calculations to avoid bias that can be caused by assumptions about beneficial energy saving by some technologies, which fluctuate depending on energy prices.

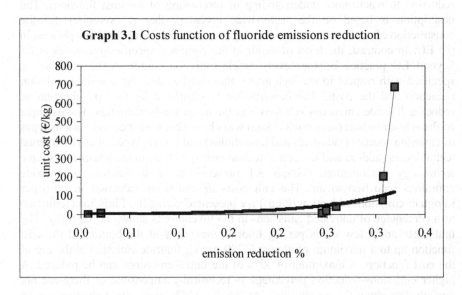

Graph 3.1 Costs function of fluoride emissions reduction

Table 3.1. Inventory of source technology combinations for fluoride emissions reduction, based on Sloof, et al., 1989 (all major sources 2 539 000 kg)

nr	Column 2 Branches	Column 3 Emissions sources	Column 4 Untreated emissions (kg/year) q_i	Column 5 Technology	Column 6 Emissions reduction kg/year e_i	Column 7 Residuals left after treatment l_i	Column 8 Demand to reduce at source % d_i	Column 9 Demanded reduction % total D_i	Column 10 Total costs €/year C_i	Column 11 Unit costs €/kg, c_i
1	Household	Old bottles	56 000	Higher recycling old glass	18 000	38 000	32.1	0.07	24 885	1.4
2	Fine ceramics	Flue gas	35 000	Dry treatment and CaO	33 000	2 000	94.3	2.0	230 415	6.9
3	Brick	Flue gas	600 000	Dry treatment and CaO	570 000	30 000	95.0	24.4	4 068 295	8.3
4	Aluminum industry	Anode preparation	298 000	Adsorption at aluin processing	12 000	286 000	4.0	24.9	230 415	19.3
5	Glass fiber industry	Flue gas	30 000	Alkaline flue gas treatment	26 000	4 000	86.0	26.0	1 152 074	44.2
6	Aluminum industry	Vapor in halls	286 000	Fluegas washing	120 000	166 000	42.0	30.7	9 216 590	77.0
7	Phosphate production	Phosphoric acid process	20 000	P-acid various measures	1 000	19 000	5.0	30.7	207 373	207.4
8	Iron and steel	Cindering	39 000	Electrofilter SO_2 washing	30 000	9 000	76.9	31.9	20 737 327	691.2
	Netherlands	Used data	1 364 000		810 000	554 000			36 403 737	Aver. 44.9

Columns 1 to 5
Column 1 shows eight source technology combinations ($N = 8$) in ascending order of unit costs that can be found in column 11. The ascending order is set in the Criteria Document on Fluoride, which is in line with the policy making that the enforcement of stricter environmental demands should start with the lowest unit costs, then the higher unit costs, and so on. Column 2 presents seven relevant branches for fluoride emissions into the air: households, fine ceramics, the brick industry, glass fiber, aluminum, phosphate, iron and steel. In a few branches there can be more emission sources; the sources are given in column 3. In this way, the inventory includes nine emission sources, e.g. aluminum production covers two emission sources: anode production (nr. 4) and evaporation in the halls of aluminum production (nr. 6). In other branches, only one type of source is specified, but in reality, a branch can cover thousands of emission sources, for example, households involved in glass recycling (nr. 1). Column 4 shows the untreated fluoride emissions going into the air. The total measured fluoride emissions going into air, water and waste in the mid-1980s was about 73 778 metric tonnes a year. Most of it was emitted into water (>58%). The total untreated fluoride going into the air from all sources was about 2 539 tonnes a year. The cost function embraces only the large emissions source that cause 1 364 tonnes of fluoride emissions a year ($Q_n = 1\ 364\ 000$ kg). The scale of emissions per source (q_i) varies from 20 to 600 tonnes a year. Column 5 presents six types of environmental technologies that can reduce fluoride emissions released into the air: 1. Glass recycling (product integrated), 2. Dry fluegas treatment with CaO (add on treatment), 3. Adsorption at aluin earth processing (add on treatment), 4. Fluegas treatment (add-on treatment), 5. Cleaner phosphoric acid process (process integrated), 6. Electrofilter (add-on treatment). These are the technologies selected by experts in the Netherlands during the 1980s projecting to comply with the demanded emissions reductions at the lowest costs in the selected sources.

Columns 6 to 9
In column 6, data on the annual scale of emissions reduction (r_i) are presented. Emission reduction is estimated on the assumption that environmental technology is applied at a source. The scale of emissions reduction varies per source. The demanded emissions-reduction percentage (d_i) at the source technology combination (i) is the scale of emissions reduction divided by the untreated emission at the source multiplied by 100. For example, the source technology combination number 1 reduces 18 tonnes of the 56 tonnes untreated fluoride emissions by avoiding production of new glass due to extra glass recycling. This reduces 32.1% of the untreated emissions at that source, whereas the combination nr. 2 can achieve 94.3% fluoride reduction at that source. The total or the cumulative emissions reduction is the sum of emissions reductions at every combination in the inventory, which are 810 tonnes of fluoride. The untreated emissions of the inventory is 2 539 tonnes, so the emissions-reduction percentage in the inventory of combination nr. 1 is 18/2 539 = 0.07%, of the combinations 1 and 2 is (18+33)/2 539 *100% = 2% and so on. If all technologies are used, 810 tonnes are reduced. This means 31.9% of the total untreated fluoride emissions released into the air can be reduced from all large sources in the Netherlands.

Columns 10 and 11
The total costs of emissions reduction (C_i) are the capital costs derived from investment plus the operational costs. For convenience, only the sum of these items is presented. The investments include civil works (like adaptation of halls) and electro-mechanic equipment (like filters). The investments are done before the operations start. The installations remain for many years; the civil works usually 20 to 25 years, the electro-mechanic equipment for 10 to 15 years. The annual costs of investments, the capital costs, are accounted for the depreciation of the investment (I) during (t) years discounted at the interest rate (r). The public investment is usually depreciated at an interest rate of 4% to 5%, the private one at 10% or more, which means at an interest of 104% to 105%, respectively 110%, or more interest. The formula for the annuity depreciation is:

$$C_I = I \frac{r}{1 - \frac{1}{r^t}} \tag{3.2}$$

Where C_I are capital costs, I investment costs, r interest rate and t is time of depreciation, for $r \geq 1$, or $r \geq 100\%$

The annual operational costs cover labor (operators, maintenance and managers), fuel and electricity, chemicals and so on. The total annual cost is the sum of capital costs and operational costs minus the revenues from sales of products (like old glass for recycling) and savings (like energy savings). The annual costs are formally:

$$C_i = C_{I_i} + O_i \tag{3.3}$$

Where C_i is total cost of emission reduction, C_{I_i} is the capital cost and O_i the operational cost at a unit i.

The annual costs of an emission source vary greatly, for example € 25 000 for combination nr.1 to € 20 373 000 for combination nr.8. There are large differences in the costs even for the same type of technology; e.g. the fluegas treatment in the glass fiber costs € 1.1 million (combination nr. 5) and in the aluminum plant € 9.2 million (combination nr. 6), but the scale of emissions reduction at the latter is larger. The cost of glass-fiber production is based on one source at one company, whereas the cost at the aluminum plant is the average cost of two sources that are two aluminum plants in the Netherlands. Some costs are based on the average costs of many sources, e.g. the net cost of glass recycling is estimated at € 25 000 (nr. 1). This is based on:

(a) Disposal costs at households calculated with regard to various ways for glass collection in municipalities and the average amount of glass disposal per household;

(b) The average transport costs per kilometer and the average distance from the disposal sources to the recycler;

(c) Difference in the production costs of new and old glass;

(d) Revenues, which equal the price of white, brown and green glass multiplied by the amount processed per category.

Column 11

The unit costs of pollution reduction (c_i) counted for every source technology combination. The unit costs are the annual costs divided by the annual emissions reduction at the source:

$$c_i = \frac{C_i}{e_i} \qquad (3.4)$$

Where c_i are unit cost, C_i total cost and e_i emissions reduction at the source technology combination i.

The total cost and scale of emission reduction are often an average of several sources so the unit costs are rather an average cost of aggregated sources per kg emissions reduction. Inter alia, the division of emissions reduction and the costs can be difficult to make, for example, in the case of various emission sources with only one outlet, like one chimney. In such cases, the costs are allocated between the sources, preferably on a mass basis.

The cost-engineering function of emissions reduction includes the unit costs at every source technology combination. It is constructed by putting the unit costs in ascending order (first, the lowest unit cost combination, then the higher unit costs and so on) and relating the unit cost to the cumulative emissions-reduction percentage. In Table 3.1 one finds eight combinations in ascending order. The lowest unit costs are at nr. 1, c_j = € 25 000/18 000 kg = € 1.4 per kg for 0.7% emissions reduction, the following is c_j= € 6.9 for 2% emission reduction and the highest is at nr. 8, which is c_j = € 2 0737 000/30 000 kg = € 691.2 per kg for 31.9% emissions reduction. The unit costs increase exponentially.

EU policy uses the average cost in the inventory as the yardstick to define the BAT. The average cost is the total cost (C) divided by total emission reduction (R) in the inventory. It should be noted that the average cost often differs substantially from the costs of individual combinations. In the case of the fluoride inventory, the average cost is € 44.9/kg (€ 36 403 737/810 000 kg). The average cost is 32 times higher than the lowest cost and 15 times lower than the highest cost. The average cost is close to the unit cost of combination nr. 5 and well below the unit cost of combination nr. 6. According to EU policy, the sources 6, 7 and 8 should be excluded from enforcement of environmental standards because the unit costs are above the average costs.

3.4 Effectiveness of Policy Making

One question is whether the EU policy based on BAT is effective. As mentioned in the previous chapter, about 80% of emissions reduction in ten to fifteen years in Europe should be demanded to approach compensation for production growth and a necessary level of environmental qualities. For this, several technologies are already available in every inventory. Table 3.2 shows selected data to analyze the effectiveness of environmental technologies: the inventories, number of technology–source combinations, number of types of environmental technologies in the inventories, emissions reduction from the largest to the smallest combination, total emissions-reduction percentage in the inventory, percentage emissions reduction at the largest combination, the lowest and the highest unit cost, the average unit cost, the percentage emissions reduction that can be achieved by enforcement of emission standards at the sources up to the average unit cost and finally the percentage emissions reduction achievable at the average unit cost in comparison to the maximum achievable (percentage reduction at average costs divided by maximum reduction percentage).

On average, one technology covers not many more than five sources in the sector-wide inventories and not more than 10 sources in the company-specific inventories. This shows that combating emissions should be tuned to the specific sources. In other words, environmental technologies are often source-specific. The demanded 80% emission reduction can be reached with the available technologies in 6 out of 19 sector-wide inventories and in 8 out of 9 company-specific inventories. This indicates that there is not a real shortage of effective technologies if the technologies are specified for the emission sources in every company. In a few sector-wide inventories effect-increasing technological development is needed because the available technologies are insufficient, notably for reduction of heavy metals released into water, phenol going into water, some volatile organics and fluoride released into the air. The demand cannot be attained solely by actions at the large sources in the inventory as in only 5 of 19 sector-wide inventories can more than 50% of total inventory reduction be attained by the technology at the largest source and in none of nine companies' specific inventories. Only SO_2 and heavy metals released into air can effectively be contained at a few of the largest sources that are electric power plants and waste incinerators. In most cases, environmental technologies should be disseminated among many sources to attain the demanded emissions reduction.

The highest unit costs are generally many times higher or lower than the average unit costs in the inventories. It is questionable whether the EU policy based on enforcing BAT only at the sources that confront unit costs below the average unit cost in the inventory is effective. Regarding the demanded emissions-reduction percentage, the EU policy is particularly effective for heavy metals, VOC to air and phosphates released into water. In most inventories, the enforcement of emission standards must embrace all emission sources to reach the target. The policy based on the BAT can cover 80% or more of emission reduction only in 8 out of 19 sector-wide inventories that are illustrative for EU policy preparation and in five out of nine company-specific inventories. This means the EU's current way of environmental policy making is not sufficiently effective for most emissions.

Likewise, the average unit cost in the inventory does not say much about the cost at individual sources because the unit costs strongly vary. The spread is large so the average costs obscure the difference between the emission sources. The spread is more than 100 times in most of the inventories and less then 10 times only in the cases of NH_3, PAH, and phenol. The spread in the inventories with a few source technology combinations is smaller, though it can be as high as 384 times, as it is for cadmium emissions reduction. A huge spread is found in the company-specific inventories than in the sector-wide inventories. This indicates that far-reaching emission reduction must consider spread in the unit costs.

These findings suggest that the available technologies are usually sufficiently effective if they are widely disseminated at emission source in companies. It means that the environmental policy is effective if all companies are involved, not only the large ones, or the ones below the average unit costs. The unit cost at emission sources increases quickly at higher emissions-reduction percentage. This contradicts the policy based on the average unit costs in a sector. To attain the demand for emission reduction that enables progress towards sustainable development while avoiding excessive costs at the sources, the unit costs at every source must be assessed. The latter means that reliable cost functions of emissions reduction must be constructed. We address the possibility of constructing such cost functions in the next sections with consideration of the scale of emissions reduction and the technological properties of the sources.

3.5 Emissions Reduction Scale and Unit Cost

Following the neoclassical theory, it should be possible to construct the cost function of emissions reduction for homogeneous inventories based on the scale of emissions reduction. If the theory holds, then policy makers can forecast the costs by using the scale factors and focus enforcement of more stringent regulations on the largest sources because these have the lowest unit costs. However, analysis of empirical data shows that this expectation is generally not valid.

Most homogeneous inventories are extracted from company-specific data. The expectation that the unit costs increase faster than the scale of emissions reduction is tested with a few homogeneous inventories, which means one type of environmental technology that can be applied in one company at more than four emission sources in that company. The scale of emissions reduction at the sources is ranked in descending order and unit costs are ranked in ascending order, as should be expected when using the neoclassical theory. Then the rank correlation (R^2) is calculated. A rank correlation between the scale and unit costs above 0.9 ($R^2 > 0.9$) is assumed to be significant. The result is that only 7 out of 27 homogeneous inventories contain a significant correlation between the unit costs and the scale of emissions reduction. Most costs cannot be forecast based on the scale. It is also checked whether the costs of some specific technologies are sensitive to the scale, but there is no specific technology with the costs that are exceptionally sensitive to the scale of emissions reduction.

Table 3.2. Effectiveness of environmental policy regarding available technologies in sector-wide inventories and company-specific inventories

	Source–technology combination	Types of technology inventory	Largest/smallest source	Per cent reduction inventory	Reduced at largest source	Lowest unit cost	Highest unit cost	Average unit cost	Attainable by average cost	Attainable: by average cost to maximum cost
A. Sector-wide inventories										
Benzene	13	8	48	66%	27%	0.9	987	193.3	65%	98%
Cadmium	4	3	10	66%	51%	150	57 603	38 121.4	33%	49%
CO_2	79	31	268	59%	7%	0.001	1.4	0.1	55%	93%
Copper	4	4	8	35%	24%	76	1 260	849.5	22%	62%
Fine dust	13	4	74	72%	10%	0.5	9.4	5.7	37%	51%
Fluoride	8	6	570	32%	70%	1.4	691	44.9	26%	81%
Phosphate	10	3	161	97%	65%	1.9	68	8.7	90%	92%
Metals to air	13	2	474	98%	77%	82	1 198	155.3	81%	83%
Metals to water	14	3	760	72%	23%	11	614	82.1	62%	86%
NH_3	10	5	60	59%	25%	1.7	14	3.2	32%	55%
NO	99	11	5 399	75%	23%	0.1	23	2.1	46%	61%
PAH	6	5	256	85%	47%	84	822	296.5	45%	54%
Phenol	7	6	6	87%	9%	0.9	53	16.6	23%	27%
Propylene	7	5	3	73%	18%	0.9	26	4.7	60%	82%
SO_2	35	7	10 231	87%	69%	0.2	3.5	0.6	64%	74%

Table 3.2. (continued)

Styrene	15	7	95	79%	16%	1.0	35	11.1	54%	69%
Toluene	20	9	398	41%	33%	0.02	461	47.9	31%	77%
VOC	47	15	90	73%	11%	0.3	7	1.9	62%	85%
Zinc	6	5	26	80%	33%	46	3 383	804.5	54%	68%
B. Company-specific inventories										
Cl in metal industry	23	9	1 040	100%	23%	1.7	656	5.0	64%	64%
NO chemical	253	10	364 027	90%	11%	0.003	1 005	2.9	73%	81%
NO electric power	97	7	346	93%	5%	0.1	99	2.3	75%	81%
NO metal	70	10	461 662	76%	21%	0.1	16 109	16.0	46%	60%
NO refineries	39	3	46 790	89%	32%	0.4	878	8.4	66%	75%
SO₂ chemical	53	6	27 327	95%	39%	0.6	2 166	3.7	79%	83%
SO₂ electric power	17	2	50	95%	12%	0.4	4.9	0.6	82%	86%
SO₂ metal	43	5	4 769	82%	33%	1.2	188	5.8	59%	72%
SO₂ refineries	17	2	473	85%	42%	1.2	31	2.2	68%	81%

One can expect that the relationship between the scale and the unit costs of a source is strong for the add-on treatment technologies because they are less dependent on the variables of the main process than the process-integrated technologies. This hypothesis is tested for two end-of-pipe technologies that can be applied at many emission sources in one branch. The technologies are selective catalytic reduction for NO_x reduction in the chemical industry, electric power plants and refineries, as well as fluegas washers for SO_2 reduction in the chemical industry and in basic metal companies (data for other inventories and technologies are not available). For each technology, the unit costs are calculated in four scale classes of emission sources in descending order: more than 400 tonnes of emissions a year, 200 tonnes to 400 tonnes, 100 tonnes to 200 tonnes and less than 100 tones. The scale of the largest sources is many times more than four times larger than the scale of the smallest ones. This is because there are many very small sources in the inventory. In between, the scale of the classes decreases roughly by half, so one can expect more than doubling of the unit costs between the classes. The unit costs are the average cost per class, that is, the total cost in a class is divided by the total emissions reduction in that class. This is shown in Table 3.3.

Table 3.3. Relation between unit costs and scale of emissions reduction

Unit costs in €/kg, scale in tonne per year, n number of source technology combinations				
scale classes	> 400	400–200	200–100	<100
NO_x emission reduction by SCR				
Chemical industry, n	7	10	13	98
Unit costs	2.3	5.9	13,8	34.1
Index unit costs	100	256	600	1482
Electric power plants, n	16	28	2	3
Unit costs	4.6	17.5	15.2	39.2
Index unit costs	100	380	330	852
Refineries, n	2	5	2	13
Unit costs	6.9	11.1	13.8	22.5
Index unit costs	100	160	200	326
SO_2 reduction by gaswashing				
Chemical industry, n	4	1	2	40
Unit costs	0.9	2.3	4.1	46.5
Index unit costs	100	255	455	517
Metal industry, n	4	4	3	25
Unit costs	3.2	18.9	9.2	22.5
Index unit costs	100	590	287	703

The results do not confirm the expectations based on the neoclassical theory about scale effects on the unit costs in cases of end-of-pipe technologies. In all cases, the unit costs of the largest emission sources are much lower than the unit costs of the smallest sources, which is in line with the theory. However, this can be explained by the extremely high unit costs of very small sources (well below 100 tonnes per year). The increase in two classes (100–200 and 200–400 tonnes per year) is well below the expected doubling of the unit costs, except for the selective catalytic reduction to reduce NO_x in the chemical industry. The unit costs of the selective catalytic reduction for NO_x reduction in the chemical industries are sensitive to the scale, whereas in the electric power plant and in the refineries are not sensitive to the scale of emissions reduction. Similarly, this was found for the unit costs of the fluegas washers for SO_2 reduction in the chemical industry and basic metal companies.

The results indicate that the scale of emission reduction is only relevant in some cases. Process variables largely determine the unit costs, almost irrespective of the scale of emissions reduction, besides the very small emission sources that have extremely high unit costs. The escalation of the costs of environmental technologies can not be generalized even in homogeneous inventories of source technology combinations because process variables are so important. Hence, it is not reliable to construct the cost functions of emissions reduction based on the scale factors, even for one type of technology at similar emission sources.

3.6 Streamlined Cost Functions

The alternative theory on the cost-engineering functions suggests that the process variables determine the unit costs, but it does not provide a theoretical basis to specify which variables are important. The question is if it is possible to construct cost functions of emissions reduction without specific knowledge about the process variables.

The answer is based on the empirical observation that cost functions of emissions reduction increase exponentially. This is demonstrated by the streamlined cost function for fluoride emissions and it is also found in other inventories. Following this observation, it is postulated that unit costs are related to the emissions-reduction percentage by logarithmic interpolation of the data between the combinations with the highest and the lowest unit cost because these two reflect the most diverging process variables. If the postulate holds, then streamlined cost functions with limited data can be constructed. The steepness of the cost line can be related to the emissions-reduction percentage, albeit it is not possible to link a specific unit cost to a specific source. In practice, streamlined cost functions have little value in policy making that is based on enforcing emissions standards in permits at emission sources. This is because the unit costs at every source cannot be established. Engineers would need to determine the main process-specific factors that specify the costs of emission reduction, which is laborious. The streamlined cost functions do have a practical value for the policies that aim to put a price on emissions because the price corresponds with the unit cost on the streamlined cost function. Hence, the streamlined cost function indicates the percentage of emissions reduction that can be attained by pricing the emissions.

We test the postulate using scenarios for the streamlined cost functions and through comparison of the streamlined cost functions with the empirical ones. Thereafter, statistical theory is used to explain the results. The logarithmic interpolations of the unit cost and the scale are done with exponents: k_c for the unit costs and k_e for the scale of emissions reduction. Both exponents are derived from the unit cost and the scale of the source technology combination. These come from the combination with the highest unit costs (n) and the combination with the lowest unit costs (1), respectively, from one with the largest and smallest scale of emissions reduction.

For k_c it is formally:

for $i + m$ with known empirical values of c_i and c_{i+m}

$$c_{i+m} = c_i \cdot e^{k_c \cdot m} \text{ ; e is Euler number (ca. 2.718)} \tag{3.5}$$

$$\frac{c_{i+m}}{c_i} = e^{k_c \cdot m} \tag{3.6}$$

$$\ln(\frac{c_{i+m}}{c_i}) = k_c \cdot m \tag{3.7}$$

$$k_c = \ln \frac{(\frac{c_{i+m}}{c_i})}{m} \tag{3.8}$$

Thus, for the combination n and 1 it is:

$$k_c = \ln \frac{(\frac{c_n}{c_1})}{n-1} \qquad \text{for cost exponent} \tag{3.9}$$

$$k_e = \ln \frac{(\frac{e_n}{e_1})}{n-1} \qquad \text{for scale exponent} \tag{3.10}$$

Then the unit costs and the scale of emissions reduction are consecutively accounted for, starting from the combination with the lowest unit cost:

$$c_{i+1} = c_i \cdot e^{k_c} \tag{3.11}$$

$$e_{i+1} = e_i \cdot e^{k_e} \tag{3.12}$$

In this estimate, the detail data are needed for two source technology combinations. The following empirical data are needed to forecast the streamlined cost functions: the scale and unit costs of the combinations with the highest and lowest unit costs (c_i and c_n, e_i and e_n), the total emission in the inventory (Q) and the number of source technology combinations (n). The streamlined cost functions are constructed based on the exponents and compared. The exponents are derived from the interpolation between the highest and the lowest unit costs combinations and from the interpolation with the exponents derived from the largest and the smallest scale combinations. In these two scenario's it is assumed that data about two source technology combinations are known but all the others are not known.

In addition, scenario's are done based on the interpolations with the known data about the scale and ranking of emission sources on cost functions. Three variants are elaborated: (1) all cost data are known and the trend is extrapolated, (2) cost data about two random combinations are known, (3) cost data about the highest and the lowest unit costs are known.

One must assess if the streamlined cost functions are accurate and reliable. The assessment is assumed to be accurate if the rank correlation between the empirical unit costs (independent variable) and the streamlined unit costs (dependent variable) is above 0.9 ($R^2 > 0.9$). A streamlined cost function is assumed to be reliably defined if the estimated total cost in an inventory deviates less than 30% from the empirical total cost in that inventory ($130\% \geq C_{emp} /C_{kc} \geq 70\%$). The calculations are illustrated in Appendix A using the fluoride example. The results of scenarios are summarized in Table 3.4. The following was found:

- The streamlined cost functions based on the exponents that are derived from the highest and the lowest unit costs are often accurate. The results are accurate even if the data about the scale of emissions reduction are not known; 17 out of 19 sector-wide and 5 out of 9 sector-wide are accurate.
- The streamlined cost functions based on the exponents derived from the smallest and largest scale of emissions reduction are often inaccurate; though these results are better for the company-specific data; 8 out of 19 sector-wide and 7 out of 9 sector-wide are accurate.
- If the scale and the ranking of emission sources are unknown, then all scenarios are unreliable. This means the total costs in the inventory based on the streamlined cost function differ too much from the empirical cost functions of emissions reduction.
- If the scale and the ranking of emission sources are known, then the streamlined cost functions are often accurate and reliable for the combinations with known lowest and highest unit costs and they are often accurate even with two random combinations, but less accurate and reliable in cases of the interpolation between the largest and smallest scale, which is not presented in Table 3.4 but it underlines poor linkage between the scale and unit costs.

The results confirm the postulate that streamlined cost functions can be accurately assessed by the logarithmic interpolation between the sources with a high unit cost and a low unit cost, but they cannot be reliably related to the emissions-reduction percentage because process variables strongly determine the costs. The logarithm extrapolation between the combinations with the highest and the lowest unit costs is in line with the empirical data. All linear interpolations and the exponential interpolations between the smallest scale and largest scale are usually not accurate and unreliable even if the data about the scale and rank are known. The low reliability of the streamlined cost functions, which means a large difference between the empirical total costs and the results of extrapolation, is mainly caused by unreliable estimates for the scale of emissions reduction. Policy makers can generally forecast the steepness of the cost functions, but not the costs at the specific emission sources in the inventory because no theoretical basis is found to relate the costs to process properties or with the scale of emissions reduction.

The streamlined cost function can be constructed with the following data: n the number of source technology combinations of the function, c_1 and c_n the unit cost of the lowest unit cost combination and highest unit cost combination and e_1 and e_n the scale of these combinations. This is needed for calculations of the costs exponent k_c and scale exponent k_e, and then cost and effects of subsequent combinations starting with the data of the first combination. The results are accurate irrespective of the spread in the unit costs. For example, they are accurate for the cost function of NO_x reduction in the chemical industry in which the spread is more than 50 000 times (the lowest is € 0.02/kg and the highest is € 1 005/kg) and NO_x reduction in the basic metal industry with the spread of more than 161 000 times (the lowest is € 0.1/kg and the highest is € 16 110/kg).

Table 3.4. Accuracy and reliability of streamlined cost functions for the 19 sector-wide and 9 company-specific inventories

Number of inventories	Accuracy: correlation between estimates and empirical unit costs $R^2 > 0.9$		Reliability: deviation below 30% compared to empirical data $130\% \geq C_{emp}/C_{kc} \geq 70\%$)	
	Sector-wide	Company-specific	Sector-wide	Company-specific
Total	19	9	19	9
Scale of emissions reduction not known				
k_c: the highest and the lowest unit cost interpolation	17	5	7	2
k_c for the largest and the smallest scale interpolation	8	7	3	4
Scale of emissions reduction is known				
$k1$ all cost data known	18	7	18	7
$k2$ two at random	17	7	10	4
$k3$ highest–lowest unit cost	18	9	16	5

3.7 Explanation of the Results

For decision making, it is important to explain why it is sufficient to know the exponents and only one source technology combination for accurate cost forecasts and then, how to find the combinations with the lowest and highest unit costs needed to calculate the exponent. The explanation is based on the statistical theory of sampling that is used in quality management. The theory is used to define the size of a product sample needed for assessment of product quality in the whole inventory. The inventory of the source technology combinations is analogous to the inventory of products.

Sampling is done to avoid laborious quality controls of all products in the inventory. A typical question is: what is the probability of deficiency of the quality in the inventory regarding the observed deficiency in the sample. The assumptions to answer the questions are that there are many products and that the deficiency is randomly distributed. The question can be answered by using the Poisson distribution. The Poisson distribution gives the probability of deficiency in an inventory by a random variable (x) that is the number of samples and a positive number (m) that is the average number of the deficiencies in the sample. It is formally:

$$P_{(x)} = e^{-m} \cdot \frac{mx}{x!}$$
(3.13)

Where e is Euler number.

If one sample is sufficient, or it is acceptable that the probability changes every sample, $x = 1$, the probability in the inventory is defined by the observed deficiency in the sample and calculated consecutively, it is one sample after another (Mood and Graybill, 1963).

$$P_{(x)} = e^{-m}$$
(3.14)

if $x = 1$

In analogy to the product quality the unit costs and the percentage of emissions reduction can be sampled in an inventory of source technology combinations on the assumptions of many randomly distributed combinations. The result is the probability function of the unit costs and per cent of emissions reduction. For example, SO_2 reduction on ships using heavy fuel, diesel or gas oil for fuel (sources) without using any environmental technology, tuning the engine or gas washing (environmental technologies) gives nine source technology combinations that are shown in Scheme 3.1. Of the nine combinations, $x = 9$, there are seven combinations for more than 50% SO_2 reduction, five for more than 90% reduction and three for more than 91% reduction, which means: $m = 7$, $m = 5$, $m = 3$. The probability of finding an additional combination is therefore: $P_{(7)} = 2.718^{-7} * 7 * 9 / 9! = 0.004 * 40\ 353\ 607 / 362\ 880 = 0.48$, for $P_{(5)} = 0.11$ and for $P_{(3)} = 0.005$. In practice one does not need to know all combinations but only the number of possible combinations, most-probable and the least-probable combination. Linking the unit costs with the probability function gives the streamlined cost function.

Scheme 3.1. Index SO_2 emission on ships, based on Krozer, et al. 2003

Shippers' options		Heavy fuel	Diesel	Gas oil
S per cent in fuel		5%	1.50%	0.50%
no technology	reduction 0%	100%	30%	10%
engine tune	reduction 20%	80%	24%	8%
gas washing	reduction 90%	10%	3%	1%

The streamlined costs indicate the probability of finding an additional unit cost as a function of emission reductions percentage. This is on the assumption that many possible source technology combinations can be found and that the costs for an additional emissions reduction increase. The theory predicts the increase of the unit costs, based on the exponent k_c and a base-combination (x_1), as function of the exponent k_e that approximates the increase of emissions-reduction percentage in an inventory. This, in turn, can be related to the increase of the unit costs as a function of increases of emissions-reduction percentage.

The question of how to find the source technology combinations with the lowest and the highest unit costs is difficult to answer using economic theories. It is instrumental to use the theory on loss prevention that is introduced in the previous chapter. On the assumption that the unit costs reflect the efforts to reduce emissions, it should be expected that the efforts are lowest at the processes with only a few inputs in the well-established production. Vice versa, the unit costs are high at the processes with many inputs and the unbalanced production. Although the explanation is plausible, it is not tested with empirical data because detailed engineering analyses are needed for this purpose.

3.8 Conclusion

Environmental policies in many countries rely on emissions standards that are enforced at every emission source to ensure that emissions reduction is attained. This policy making has invoked development and implementation of effective technologies. Many emissions can already be reduced by 80% to as much as 95% with available technologies, but it is costly. To avoid the unreasonably high costs, policy makers make inventories of emission sources with technologies to control them and estimate the average costs in an inventory based on case studies at a few selected companies. Estimates with autoritative databases and models reveal that such a policy is insufficiently effective and efficient. If the demands are only enforced at the large sources, or at the sources below the estimated average costs the aim of sustainable development cannot be attained because of insufficient emission reduction. The inefficiency in policy making based on the average costs in an inventory is that some companies are confronted with much higher than average costs, because the costs increase exponentially as a function of emissions-reduction percentage. The spread of the unit costs between companies is so large that an average cost obscures the reality but there is no adequate method to assess unit costs at every source during policy preparation.

One possibility is to escalate the costs based on the unit costs at small and large emission sources. However, escalation does not provide reliable results even for homogeneous inventories with one environmental technology at one company with a few emission sources. The escalation is also unreliable for end-of-pipe technologies. The costs largely depend on process variables. Therefore, the costs functions of emissions reduction should be related to process variables of the sources. Yet there is still little basis to classify the variables in relation to the costs.

An alternative in policy making is to take for granted that the unit costs at emission sources cannot be assessed. The observation that the costs largely depend on process variables provides a starting point in the definition of streamlined cost functions that can be used in policy making. The streamlined cost functions indicate the increase of the unit costs as a function of emissions-reduction percentage without specification of the unit cost and scale at emission sources. The streamlined cost functions are defined with the empirical data about the source technology combinations with very high and very low unit costs, data about total emissions in the inventory and number of combinations. The cost and emission exponents can be calculated by the interpolations between these two combinations and the subsequent unit costs and emissions percentage are accounted consecutively starting from the lowest unit cost combination. It is formally:

$$k_c = \ln \frac{(\frac{c_n}{c_1})}{(n-1)}$$

$$k_e = \ln \frac{(\frac{e_n}{e_1})}{(n-1)}$$

Hence,

$$c_{i+1} = c_i \cdot e^{k_c}$$

$$e_{i+1} = e_i \cdot e^{k_e}$$

The steepness of the cost function can be accurately forecast ($R^2 > 0.9$) for 27 out of 28 inventories. The limitation is that the total costs in the inventories are not reliably assessed, because the scale of emissions reduction cannot be well forecast. The explanation of the results with statistical theory is that the steepness of the streamlined cost functions indicates the probability to find consecutively unit cost as function of emissions-reduction percentage in an inventory of source technology combinations. The theory on loss prevention suggests that the lowest unit costs are found in the processes with the least inputs and mature processes and vice versa, the highest costs in the processes with many inputs and low operational experience.

The results have several implications for policy making. Firstly, the policy cannot foresee the costs of enforcement at individual emission sources because there is no sound theoretical basis for such estimates. Secondly, the enforcement aiming at sustainable development in the EU must move far beyond the BAT to be effective. Third, the exponential cost function of emissions reduction means that the estimate of the average costs in a branch, even a large breadth of the average cost, does not say much about the real costs at the sources. Fourthly, it is possible to forecast the streamlined cost functions with the empirical data about only two source technology combinations. The streamlined cost functions are useful to define the price of emissions needed to attain a targeted emissions-reduction percentage. This suggests that pricing policies can be easier to prepare than policies based on emissions standards. Fifth, it is erroneous to assume that large sources have low unit costs due to economies of scale because process variables largely determine the costs. Effort should be put into understanding how process variables influence the costs of emissions reduction and into classification of processes with respect to costs and effects of environmental technologies.

4

Projecting Innovation Costs and Benefits

If to progress with environmental policy at a reasonable costs a question is: Will expenditure on environmental innovations provide a social benefit? The answer is relevant to underpin the funding of technology development. It must be given without knowing the costs and effects of innovations during use at the moment of decision making about technology development. Streamlined cost functions are used to determine if it is more attractive to invest in technology development (innovation) or to improve an available technology during its use (adaptation).

4.1 Adaptations and Innovations

Decision makers consider many proposals for new environmental technologies that can substitute for available past technologies. The usual deliberation is which proposals for innovations provide a large social benefit as measured by the costs of emissions reduction. At the same time, it is important to be aware of alternatives that are adaptations of available technologies with gradual cost savings but without expensive technology development. The expected cost savings due to the innovation should be substantial to motivate investment in research, development and manufacturing that ultimately enable profitable sales and the sales can be expected if the emission source can reduce the costs by using a new technology instead of adapting available technologies. Hence, the expectation of large cost savings at emission sources determines whether the innovations are attractive. The present value of the expected cost savings at emission sources plus the present value of the expected profit to be made by innovators provide an innovation rent. The sum of the innovation rents in an inventory of emission sources is called the surplus of innovation rent. The surplus approximates the social benefit of environmental innovations.

The issue is addressed of how to estimate the possibilities of cost savings at emission sources without knowing polluting companies' costs for using the technology. Reliable data at the moment of decision making about technology development can only be found on the costs and effects of available technologies at a few emission sources because polluters have no interest in providing such information. Such data are only sufficient to define the streamlined cost functions as shown in Chapter 3. These streamlined cost functions are used in this chapter to assess the social benefit of innovations in the inventories introduced in the previous Chapter one. Regarding the streamlined cost functions, a simpified deliberation on developing an environmental technology is illustrated in Figure 4.1. The figure represents an emission source with two technologies. The choice is whether to develop an alternative for cheap technology AB that can attain 30% emissions reduction, or to opt for an alternative for the more costly one, EF, that can reduce in addition 10% of emissions. If both alternatives can save 50% of the unit costs, the total savings are: AC times the scale of emissions reduction AB compared to EG time the scale of emissions reduction EF. The total expected cost savings are presented by the areas ABCD and EFGH, respectively.

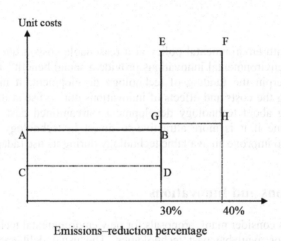

Figure 4.1. Choice of environmental policy regarding innovations

Innovations at low unit cost sources cannot save much; this only happens at large-scale emission sources. For an equal scale of emissions reduction, the saving at high unit cost sources should be larger (theoretically underpinned in Krozer, 2002:91–92). Regarding the findings in Chapter 3 that the scale and unit costs are hardly related; the large innovation rents, therefore, should be expected at the high unit cost sources. Following this argumentation, it is postulated that innovations provide a large cost saving for the steeply increasing, streamlined cost functions and small cost saving for the flat streamlined cost functions. If the postulate holds, innovating is attractive for steep cost functions. The postulate is tested that innovations provide the largest cost saving in inventories with largest cost exponents. This enables policy makers to focus efforts on innovations with data about costs and the effects of available technologies. Then the social benefit of environmental innovations for the Dutch situation is assessed.

4.2 The Attractiveness of Environmental Innovations

In line with the theory above we assume that polluting companies focus innovative efforts at the high unit costs emissions source and rely on adaptations at the low unit costs emission sources and test if the assumption holds. It is assessed whether innovations save costs at high unit cost sources or at low unit cost sources. The expectation is that the larger the cost exponent, the bigger the savings that can be reached through innovations at high unit cost sources. Figure 4.2 is a graphical presentation of the simulations. The unit costs are shown vertically and the emissions-reduction percentage horizontally. The streamlined cost function is shown by the line c_e, the simulation of adaptations by the line c_a and the simulation of innovation by the line cri. The cost savings c_s are represented by a downward shift of the streamlined cost function. The areas between c_e and c_a respectively c_e and c_i equal the costs savings. For adaptations, the unit costs at the top of cost functions do not change, however they do change at the bottom. The opposite is true for innovations: unit costs at the bottom do not change, but they do change at the top and the streamlined cost functions flatten.

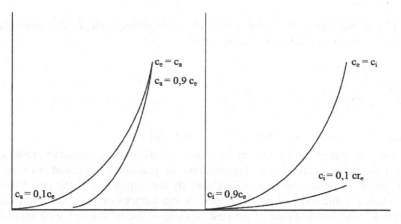

Simulation of adaptations   Simulation of innovations

Unit costs

$c_e = c_a$

$c_a = 0,9\ c_e$

$c_e = c_i$

$c_i = 0,1\ cr_e$

$c_a = 0,1c_e$

$c_i = 0,9c_e$

Emissions-reduction percentage   Emissions-reduction percentage

Figure 4.2. Presentation of simulated adaptations and innovations

The following numbers are used in the simulation. For the adaptations, the largest innovation rent is 90% at the bottom of the streamlined cost function, decreasing linearly down to 0% at the top. For the innovations, there is a 90% innovation rent at the top that decreases linearly to 0% at the bottom, whereas the bottom remains unchanged. The total cost in the simulated inventory is then divided by the empirical total costs in the inventory to calculate the costs savings. The assumptions are not far from reality. The assumed average costs decrease in an inventory is 50%, which implies an average annual savings of 3.5% within 15 years of use. Such cost savings are in line with the annual, average productivity increase in industries and it is substantially lower than the productivity increase of environmental technologies assessed in Chapter 5.

The simulation of the innovations described above can be defined by two series: the empirical data $c_{e1}, c_{e2}, .. c_{en}$ and the innovations $c_{i1}, c_{i2}, .. c_{in}$. The relation between these two is defined by a coefficient a.

$$c_{ii} = a_i \cdot c_{ei} \qquad (4.1)$$

for

$$c_{i1} = (1 - a_{i1}) \cdot c_{e1} \qquad (4.2)$$

$$a_{i1} = 0$$

and

$$c_{in} = (1 - a_{in}) \cdot c_{en} \qquad (4.3)$$

$$a_{in} = 0.9$$

$$a_{ii} = \frac{(a_n - a_1)}{(n - 1)} \cdot (i - 1) \qquad (4.4)$$

$$c_{tot} = c_{e1} - c_{i1} + c_{e2} - c_{i2} + ... + c_{in} - c_{en} \qquad (4.5)$$

or

$$c_{tot} = \sum_{i=1}^{n} (c_e - c_i) \qquad (4.6)$$

The attractiveness α_t is the index of costs savings due to the innovations divided by the empirical data in an inventory.

$$a_i = \frac{\sum_{i=1}^{n} (c_e - c_i)}{\sum_{i=1}^{n} c_e} \qquad (4.7)$$

Mutatis mutandis for adaptations but $a_{a1} = 0.9$ and $a_{an} = 0$

The example of the simulation for fluoride emissions and numerical results for all inventories are presented in Appendix B. The results are presented in Graph 4.1 and Graph 4.2. The graphs show horizontally the exponents of the cost function and vertically the indexed costs savings, or the attractiveness of the innovations. Graph 4.1 shows the spread of the costs savings in the inventories with small cost exponents. However, if the inventories are grouped in six clusters in ascending order of cost exponents the spread becomes smaller and the correlations between the attractiveness of innovations and cost exponent are high. This is shown in Graph 4.2. The correlation between the costs savings and the cost exponents in the clusters is significant for the adaptations ($R^2 = -0.92$) and for the innovations ($R^2 = 0.91$). The costs savings made through the adaptations decreases and the costs savings related to the innovation increases at larger cost exponents. The turning point is found at the cost exponent 0.24 ($k_c = 0.24$). Innovations are usually more attractive than adaptations for streamlined cost functions with cost exponents larger than 0.24. Sensitivity analyses with a maximum of 10% instead of a maximum of 90% costs savings for innovations and adaptations do not change the results but the turning point in favor of innovation moves from $k_c = 0.24$ to $k_c = 0.36$. The results confirm the postulate that environmental innovations are attractive for the cost function with large-cost exponents and that adaptations offer advantage for cost functions with small-cost exponents.

The findings can be explained. For an inventory of emission sources, the number of available technological alternatives determines the cost exponent. If many technological alternatives are available in an inventory, the probability of finding a new one that decreases the unit cost at that source is low. Conversely, if there are only a few available technological alternatives in an inventory, a steep, streamlined cost function is found and the high probability of finding a lower cost alternative should be expected. In this way, attractiveness of the innovations compared to the available technologies can be estimated.

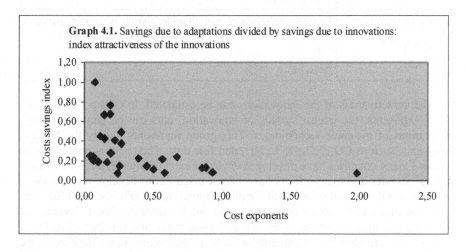

Graph 4.1. Savings due to adaptations divided by savings due to innovations: index attractiveness of the innovations

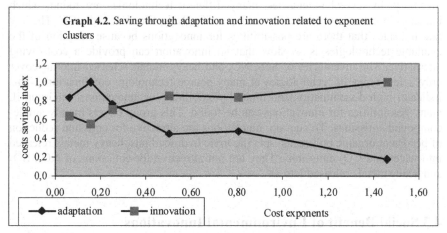

Graph 4.2. Saving through adaptation and innovation related to exponent clusters

The results provide a guideline to identify areas of policy making for innovations that can provide a large social benefit due to the high surplus of innovation rents. The innovations are attractive in inventories of source technology combinations with streamlined cost functions described by a large cost exponent, roughly the cost exponent above 0.24 ($k_c = 0.24 \pm 20\%$). Based on this criterion, the inventories are grouped with respect to the innovative possibilities. Scheme 4.1 relates the areas of environmental policy to cost exponents.

Scheme 4.1. Areas of environmental policy in relation to cost exponents (company-specific inventories are italicised)

Group I, Total 16 Small exponent $k_c < 0.24$	CO_2, SO_2, Metals released into air, NH_3, NO_x, fine dust, VOC *NO_x chemical, NO_x electrical power, NO_x basic metal, NO_x refineries, SO_2 chemical, SO_2 electrical power, SO_2 basic metal, SO_2 refineries*
Group II, Total 12 Large exponent $k_c > 0.24$	Benzene, cadmium, phenol, fluoride, phosphate, copper, metals released into water, PAHs, propylene-oxide, styrene, toluene, zinc *Cl-metal*

The attractiveness of the innovations can be estimated. In Group I comprising small exponents, the largest surplus of innovation rents can be attained through adaptations of available technologies. This group represents mainly the "bulky" emissions, such as CO_2, SO_2, NO_x, NH_3 and fine dust. The result is not surprising because many countries have developed environmental policies in these fields over several decades and many efforts have been carried out to develop and introduce environmental technologies. This group, however, should be divided into subgroups I.a and I.b based on dissemination of the technologies in inventories. The dissemination is estimated by the following ratio: the number of different technologies divided by the number of source technology combinations in the inventory. In several inventories, many different technologies are found, which indicates low dissemination. This holds for CO_2, NH_3, fine dust and VOC. The low rate indicates that there are possibilities for innovations because adoption of the available technologies is so slow that an innovation can provide a cost-saving alternative. The inventories of SO_2, NO_x and heavy metals released into air cover only a few types of technologies at many source technology combinations. This indicates high dissemination and limited possibilities for innovations. In Group II, many possibilities for innovations can be found. This group covers mainly single-compound emissions. To sum up, innovations are attractive for emission reduction of persistent organic pollutants, specific toxic hydrocarbons, heavy metals released into water and CO_2 emissions. They are not attractive for emissions of NO_x, SO_2 and heavy metals released into air.

4.3 Social Benefit of Environmental Innovations

The question follows: Does the social benefit of environmental innovations outweigh the costs of technology development? The data are from the Netherlands, but the method can be used for other countries, regions and sectors. It is a scenario on the social costs and benefits of environmental innovations based on the cost functions of emissions reduction introduced in Chapter 3. The assessment of the social benefits of environmental innovations follows mainstream innovation theory. Hence, we briefly introduce some ideas about the role of R&D in innovation, success factors in the innovation process and in dissemination, followed by assumptions about environmental technology in the Netherlands that are used to construct the scenario.

In the mainstream view, the innovation process usually starts with research and ends with the profitable sales of new technology. The research enables the development of a commercial result. The commercial result of R&D is a patent that is protection granted to an invention. The patent gives the owner the right to manufacture and sell products or to sell the right of development as a license. A newly manufactured product is usually demonstrated in use, for example in a production plant, which is called an innovation. Efforts in the innovation are often called R&DD, which includes demonstrations of the technology use. The sales of the innovation are called diffusion. The sales must cover all of the costs and risks of the innovator in order to make a profit. The costs of R&DD are considered an investment, because the expenditures are fixed over many years, independent of the sales volume. The products can be sold if they benefit the user. This means that the users' investment cost in the innovation must be lower than the savings during life cycle of use. The results of the investment decisions of the innovators and the users are uncertain at the moment of the decision making. Regarding the uncertainty, the sales and users' savings are discounted at an interest rate that reflects the risks of investment. That is, a higher interest rate is used for risky investments. The basics of innovation theory are widely accepted, albeit it is emphasized that the innovation process is not linear. Rather, it includes many interactions between know-how and market demands. The interactions determine the direction and speed of the processes. It is also pinpointed that much know-how in innovation processes is not formalized in R&D but instead is generated by various institutions, skills and cultural tradition, called tacit knowledge.

Innovators gain first-mover advantages but they face the problem of uncertainty about the results of efforts. Even the most promising piece of science or technology is no guarantee of a successful R&D result. A successful R&D result, like a patent, does not imply that the patented technology is going to be developed or that the patent holder manufactures, owns and sells the product. Only a small portion of R&D provides a patent. A few patents out of many provide a saleable product and only a few new products become profitable after some years of sales. It is often not the inventor or patent holder that becomes a successful entrepreneur and gains the profit, but rather an entrepreneur who is able to obtain a license to manufacture and sell the product (Rosenberg, 1982a). Innovations are usually measured by patents, but it is widely acknowledged that the measurements are imperfect because there is no direct relation between patents and innovations. The data on patents is only the best approximation of innovations (Grilliches, 1994). The uncertainties about R&D output and the first-mover advantages surrounding use of R&D for innovations are illustrated for organic chemistry products such as styrene, phenols, vinyl chlorides, acrylonitrile, ortho- and para-xylene, isoprene and cyclohexane. Most of these products were discovered by the end of 19th century and were applied decades after the invention, e.g. after 47 years for isoprene and even after 77 years for vinyl chloride. Not a single inventor became the innovator who successfully introduced the product. The innovators, however, gained a profit that was many times the original R&D expenditure and remained successful for decades. Though success attracted many competitors who entered the market four and a half to seven years later, resulting in price reductions on the products, all the innovators have remained market leaders for more than 50 years. First-mover advantages, therefore, are important. First-mover advantages are based on patents, as well as listening to clients' demands, lowering product costs through process innovations after the start of production, economies of scale due to product standardization and experience in production and sales (Stobaugh, 1988). Experiences in the fine-chemical industry indicate that the period between inventions and innovation, called the lead time to market, is shortened and that product life cycles have become shorter. The life cycle of products is cut roughly in half every ten years (Ganguly, 1999).

The success factors of R&D are connected with human capacities and economic conditions. An outstanding review of the human capacities can be found in Andreasen (2005) and this kind of factors is not discussed further. The economic kind of success factors in technology development change during various stages of the innovation process. In the R&D phase, the key factor is supposed to be creativity, which is fostered by conditions such as freedom for handling and valuation of innovative staff behavior. R&D results are used for commercialization by managers and financiers who are usually ignorant about the specific field of innovation, which causes funding traditions, opinions of competing experts and personal interests to prevail in decision making about commercialization. Commercial failures of positively assessed R&D output and vice versa, successes emerging from negatively assessed R&D results are no exception. Entry of the new technologies that substitute available ones are usually unforeseen because such disruptive technologies emerge in specific social networks (market niches) and expand quickly as the technology is adapted to many applications (Ekvall, 1991; Christensen, 2000).

Diffusion on the professional markets such as the market of environmental technologies is generally driven by economic advantages that are mostly gained through lower life-cycle costs and better performance. They are less affected by other factors such as promotion, distribution and services after sales. The cost performance ratio is expected to be decisive in diffusion and innovations must compete with cost-saving adaptations of available technologies (Tushman and Anderson, 1987; Cooper and Kleinschmidt, 1991). Diffusion is usually measured by the number of sales (rate of diffusion) and by the value of sales in time (the speed of diffusion). The value of sales behaves as an S function, a so-called logistic function. Units sales are slow initially because many customers hesitate to buy, then they speed up as the innovation becomes well known. Finally, they slow down again because of market saturation. The price per unit is initially high, but it decreases because competitors copy the innovation (mimicry) and because new competing products are developed (entrants). Hence, sales expand but flatten after some time and ultimately saturate.[3] Methods are developed to assess diffusion, such as multi-criteria techniques to assess innovations' attractiveness (advantages) versus impediments (risks), analytical methods to estimate the value and risks of the products based on marketing studies and market-share scenarios based on the competitive advantages of innovations and others (Van Beek, 1997; Beije, 1998).

Many issues involved in the innovation process also hold for environmental technologies but environmental innovators confront the so-called "double externality". This means that the competitors can accrue the results of R&D at low costs (spill over), which is found on all markets. In addition, it is uncertain whether the clients, the polluting companies, will buy the new environmental technology because they consider emissions as external to production and have no interest in reducing emissions unless they are demanded to do so by policy makers (Jaffe et al., 2005). In effect, environmental innovators are rarely the polluting industries, but rather universities, public institutes and suppliers of environmental technologies. For example, in the Netherlands, expenditures on environmental R&D fluctuate annually between 5% and 8% of the costs of pollution controls. These R&D expenditures grew until the early 1990s but thereafter declined to about € 120 million a year (in 1980 prices). About 65% of it is for technology development and the remaining part is mainly impact assessment. Almost all impact assessment studies are financed from public funds and more than 60% of environmental R&D for technology development is covered by the public sector. The remaining 35% to 40% of all environmental R&D for technology development is covered by private companies. About 80% of all environmental R&D is funded by public policy.

The expenditures in environmental innovations cover more activities than R&D. These are: procurement of equipment, research, industrial design, licenses, marketing and staff training (Klomp and Pronk, 1998). In the industry, about 56% of expenditures on innovations are research and industrial design, which is R&D. If we assume that this percentage also holds for environmental technology, the annual investment in environmental innovations by industries is almost twice the amount of environmental R&D. Environmental innovations are also created by services such as engineering companies. R&D in services is small compared to the industries because about 93% of expenditures on environmental innovations are for technology procurement. Hence, total R&D expenditure on environmental technology made by industries and services together is slightly below 50% of the total expenditures in R&DD on environmental innovations. The investments on environmental R&DD in the Netherlands are, on average, around € 250 million a year. R&D costs per patent in environmental technology have been assessed for the period 1979 to 1984, showing R&D costs of € 2.4 million to € 5.3 million per

---

[3] Logistic function (in population biology) describes organism growth in a closed system with the Lottka Volterra model: $\log(\frac{n_{(t)}}{N - n_{(t)}}) = a + b_{(t)}$ with $n_{(t)}$ adopter, N population, a = constant, $B_{(t)}$ = chance of adoption (speed of diffusion). Mansfield (1971) explained the diffusion by the number of years of diffusion, the cost advantage for users and the size of investment in the innovation use. This pattern is found in many innovations (Fallen, 1983).

patent, in 1980 prices (Van Driel and Krozer, 1987). Therefore, roughly € 6 million to € 10 million must be spent to demonstrate an environmental invention. However, it is unknown how many patents are actually demonstrated, how much it costs to manufacture a final product for use that is an environmental innovation and if more R&DD is needed to develop more effective and efficient technologies now that the mid-1980s have passed. The latter is not always needed because innovators successfully transfer know-how from one field to another. We are also unaware of studies on lead time to market in environmental innovations. The patent analysis mentioned above, suggests that the lead time between patent acceptance and application of the new technology based on the patent is somewhere between 7 and 12 years, but the R&D period before patenting could be longer (Krozer, 1989).

Studies on diffusion show that cost effectiveness (or unit costs) is the determining factor of success for the rate of diffusion of environmental technology. This is the most widely used cost performance ratio in environmental technology. The speed of diffusion is largely determined by the stringency of environmental policies. Most studies emphasize that not all companies implement innovations at all emission sources. Based on the studies, it can be estimated that environmental innovations attain 15% to 50% market share in 15 to 20 years in areas of policy making with many technological alternatives, such as insulation, VOC paint, membrane for heavy metals released into water and biological treatment. A higher market share of innovations is attained in areas with few alternatives (Brezet, 1994; Kemp, 1995). Methods to project diffusion are basically experts' opinions about technological options that can meet future environmental demands (back casting) and scenarios for environmental policies with possible diffusion of new technologies (Coopers & Lybrand, 1992; Benhaïm and Schembri, 1996; Weterings et al., 1997; Ruttan, 2002).

The scenario method is used to assess possible social benefit from environmental innovations. The assessment done for the Dutch situation is applicable to other situations. The purpose is to indicate the social benefits of environmental innovations taking into consideration the interests of technology developers (innovators) and users of new technologies (polluting companies). The innovators' perspectives and those of the polluting companies differ. An environmental policy is considered that aims to reduce emission based on implementation of all available technologies that are found in the inventories described in Chapter 3 within 15 years but the polluting companies can choose better technologies. Regarding such policy making, innovators will only invest in technology development if they expect sufficiently large investments at emission sources due to stricter environmental policy.

The expected investments in innovations compared to adaptations are based on the innovations' attractiveness, which is assessed in the previous section. The expected sales of the innovations are function of the investment at emission sources I, corrected by the estimate of the attractiveness of the innovation. The attractiveness of the innovations α is based on the index savings through innovations compared to the savings through adaptations, which is estimated in the previous section. The perspective of the polluting companies is focused on the cost savings β made through use of the innovations. The assumption is that the polluting companies buy and use the innovations only if they can save annual costs C during the years of the life cycle t_l. However, considering that the life-cycle cost of innovations is unknown at the moment of decision making, the arbitrarily expected cost savings are 2% a year, in line with long term economic growth. The average life cycle of 21 years is assumed, based on 60% civil depreciated in about 25 years and 40% mechanical depreciated in about 15 years. This implies that the polluting companies can save in total 35% by using the innovations compared to using only the available technologies, which is not an excessive figure. It is assumed that the innovators and the polluting industries discount the sales of innovations, respectively, the cost savings by 10% a year ($r = 1.1$ or 110%). The accounting is made for every inventory separately. The sum of the discounted sales of innovators indicates the surplus of the innovators S_i, the sum of the savings in the life cycle of use indicates the surplus of the emission sources S_e. The sum of these two provides the total surplus of the environmental innovations. The total surplus can be compared with the investment in R&DD $I_{R\&DD}$ to indicate the social benefits S_b of the innovations.

For the assessment, the investment costs I respectively the annual costs C of the source technology combinations in the inventories are distributed in fifteen years.

$$t = k_j f(n_j) \tag{4.8}$$

$$I_i = fI(t) \tag{4.9}$$

$$C_i = fI(t) \tag{4.10}$$

Where t is 15 years, n is the number of source technology combinations in a inventory and k is the distribution factor that relates the number of combinations and the number of years, j is inventory of source–technology combinations

Sales of the innovator $V(t)$ is a function of investments at emission sources $I(t)$ corrected for the attractiveness of innovations in an inventory α $(\alpha<1)$.

Hence,

$$V(t)_j = a_i \cdot fI(t)_j \tag{4.11}$$

Which is discounted to assess the innovators surplus S_i in an inventory j

$$S_j(t) = \frac{V(t)_j}{r^t} \tag{4.12}$$

For $r \geq 1$ and summed up for all inventories

$$S_{tot} = \sum_{j=1}^{n} (s_i j) \tag{4.13}$$

For the emission sources, the total costs of emission reduction $C(t)$ are the sum of capital costs $C_I(t)$ and operational costs $O(t)$. The costs savings at emission sources S_e in time is a function of the total costs of emission reduction, corrected for the annual savings factor β (<1) due to use of the innovations in the life cycle t_l. The total cost is the sum of the capital and operational costs

$$C_i = C_{Ii} + O_i \tag{4.14}$$

The savings in time in an inventory i

$$S_e(t) = f[(C(t) - \beta^{t_i}) \cdot C(t)] \tag{4.15}$$

This is discounted for an inventory

$$S_e = \frac{S_e(t)}{r^t} \qquad (4.16)$$

For $r \geq 1$ and summed up for all inventories

$$S_e = \sum_{i=1}^{n} (S_e j) \qquad (4.17)$$

The social benefit is

$$S_b = S_i + S_e - I_{R\&DD} \qquad (4.18)$$

Example of the estimates is presented for fluoride emission. Columns 1 to 5 show the empirical data: years of implementation, emissions reduction, investments, unit costs and annual costs. Columns 6 to 8 show the sales possibilities based on distribution of the empirical investment data in 15 years of implementation, followed by correction for the attractiveness of innovations that is 73% of all investments and the present value of sales based on a 10% interest rate, which is 110% interest. Columns 9 to 12 show the cost savings made through use of innovations in the life cycle: the annual costs, life-cycle cost of the technology implemented that year, cost savings in the life cycle due to the annual 2% savings and finally the present value of the cost savings based on a 10% interest rate. In Appendix B, estimates for other inventories of source technologies' emissions are presented.

Table 4.1 sums up the results: the total potential sales, the savings and the sales undiscounted including the technologies for CO_2 emissions reduction and excluding these technologies as well as the surplus of innovation rents that is the present value of the savings and the sales including and excluding the CO_2-reducing technologies. Innovators' perspectives regarding investments and polluting companies' perspective regarding potential cost savings in the life cycle are presented separately and compared with the costs of R&DD.

Table 4.1. Possibilities for environmental innovations in the Netherlands; innovation rent equals savings, social benefit equals present value of the savings

€ billion	Total investments and the life-cycle costs	Sales of innovations and savings in the life cycle due to use of the innovations		Surplus of innovation rents	
		All inventories	excluding CO_2	All inventories	excluding CO_2
Perspective of innovators	76.9	50.1	14.6	5.7	2.0
Perspective of polluting companies	202.4	94.9	52.9	24.8	12.7
Total surplus				30.5	14.7
Total R&DD	3.8	Social benefit after R&DD		26.7	10.9

The social benefit of innovations outweighs by far the innovative expenditures. From the innovators perspective, the sales possibilities, due to the investments in environmental innovations, outweigh the costs of R&DD if the investments in new CO_2-reducing technologies are included. However, if these technologies are excluded the market for environmental innovations from the innovators' perspective is too small for the present level of R&DD. The decreasing R&D on environmental technology is, therefore, not surprising because the innovators confront lower demands for environmental technologies. The users that er the polluting companies, however, would benefit a lot from more environmental innovations. The social benefits of environmental innovations, indicated by the savings gained during the life cycle of technologies, are large and outweigh many times the costs of R&DD. A sensitivity analysis indicates that even a minor cost saving in the life cycle of products (slightly above 0.5% a year) outweighs the costs of R&DD. The assessment has two major implications. Firstly, R&DD expenditures can be expanded regarding the necessary investments to reduce emissions, particularly to reduce CO_2 emissions. Secondly, R&DD expenditures can be substantially expanded when the innovations contribute to cost savings during the life cycle of technologies. This means that the development of environmental technologies is restricted by imperfect knowledge about the effects of new technologies during use such as cost savings. Imperfect forecasting of life-cycle costs impedes environmental innovations causing, unintentionally, unnecessarily high costs of emissions reduction by polluting industries.

Table 4.2. Assessment of sales and savings in € mln due to innovations for fluoride emissions reduction, interest 10%, attractiveness 73% of investments, annual savings 2%

Years	Empirical data				Sales possibilities based on investment			Cost savings based on life-cycle cost			
	Emission in kg	Investment	€/kg	Total costs	Investments in 15 years	Innovations (73%)	Present value	Annual Costs	Life-cycle cost	Saving life cycle	Present value
1	18 000	0	1.4	0.02	0.0	0.0	0.0	0.02	0.5	0.1	0.0
2	33 000	1	7.0	0.2	0.7	0.5	0.4	0.04	0.8	0.2	0.0
3	570 000	14	8.1	4.6	0.9	0.7	0.5	0.06	1.2	0.3	0.1
4	12 000	2	19.2	0.2	1.3	0.9	0.6	0.09	1.8	0.5	0.1
5	26 000	2	44.3	1	1.7	1.3	0.8	0.1	2.8	0.7	0.2
6	120 000	46	76.8	9	2.3	1.7	1.0	0.2	4.3	1.1	0.3
7	1 000	1	207.4	0	3.1	2.3	1.2	0.3	6.5	1.7	0.4
8	30 000	69	691.2	21	4.3	3.1	1.5	0.5	9.9	2.6	0.6
9	810 000	134		36	5.8	4.2	1.8	0.7	15.1	4.0	0.9
10					7.8	5.7	2.2	1.1	23.1	6.0	1.4
11					10.6	7.7	2.7	1.7	35.1	9.2	2.2
12					14.3	10.5	3.3	2.5	53.5	14.0	3.3
13					19.4	14.2	4.1	3.9	81.5	21.3	5.1
14					26.3	19.2	5.1	5.9	124.2	32.5	7.8
15					35.6	26.0	6.2	19.3	405.0	105.9	25.3
Total		134		36	134	98	31	36	766	200	48

4.4 Conclusion

Policy makers must face the challenge of assessing whether expenditures on development of new environmental technologies provide a social benefit. It is a challenge because investments in research, development and the demonstration of a new environmental technology (R&DD) face uncertain sales of innovations in comparison with an alternative that is an adaptation of available technologies from the past. Moreover, the cost savings that can be achieved by using environmental innovations during the life cycle of technologies are not known. Decisions about the development of environmental technologies, therefore, are biased because of the unknown social benefit of selling and using the innovations.

To underpin decision making, the postulate is tested that the innovations are most attractive in cases of steep cost functions of emissions reduction because there are only a few low-cost alternatives in an inventory of source technology combinations, whereas the adaptations are most attractive in cases of flat cost functions because many alternatives are available. This is tested with 28 streamlined cost functions that are elaborated in Chapter 3. The postulate is confirmed. It is found that innovations are usually attractive for streamlined cost functions with the cost exponent above 0.24 ($k_c > 0.24$), whereas adaptations are usually more attractive for cost functions with smaller exponents; sensitivity analyses show that the outcome is robust. The most attractive areas for innovations are: CO_2 reduction, emissions of hazardous volatile compounds and heavy metals released into water. Considering the attractiveness of innovations in various areas of environmental policy, it is assessed if a higher level of R&DD than the present one can be justified and whether the implementation of environmental innovations provides a social benefit.

The assessment is done for the situation in the Netherlands but the scenario method has broader applications. The social benefit is approximated by the present value of the surplus of innovation rents, which means cost savings through innovation in comparison with use of available technologies. Two perspectives are addressed. The first perspective is of the innovator seeking a sufficiently large market for sales of the innovations. It is found that the surplus of innovation rents that can be attained through strict environmental policy justifies higher R&DD expenditures than the present ones. This is particularly true with regard to CO_2 emissions reduction using renewable energy technologies. However, the R&DD expenditures are not attractive for innovators in case of low demands for CO_2 emissions reduction. The second perspective is one of the polluting industries aiming to save costs related to emissions reduction through use of the technologies. The finding is that the innovation's social benefit is large, even in cases of low annual cost savings in the life cycle, which justifies greater efforts in environmental innovations. The polluting industries can benefit a lot from the cost saving environmental innovations.

The results have several implications for innovators and policy makers. Firstly, the exponential increase of the cost function of emissions reduction implies that opportunities for sales of innovations can be found at sources that confront costly available technologies. That is, the innovators can find market niches even for costly innovations. Secondly, innovations are most attractive for the steep cost functions. This implies that policy makers can adjust R&D funds to the potentially most attractive areas solely using data about available technologies. Thirdly, due to strict policy, the growing market for environmental technologies justifies a higher than present level of expenditures on environmental innovations. Finally, development of environmental technologies falls well below a level needed to maximize social benefit due to cost savings during the life cycle of new environmental technologies. Uncertainty about the future benefits of innovations during use causes unnecessarily high costs for emissions reduction.

5

Environmental Policy and Technological Progress

Many warn that strict environmental policy causes high costs in industries without contributing to business results and has severe impacts on productivity. Other scholars bring a nuance to this, arguing that companies can reduce costs by economies of scale and innovations with positive, though unintended, positive effects on industrial efficiency. A few comprehensive literature reviews (Withagen 1999, Jaffe et al., 2002) cover the debate. After a summary of the positions, this chapter uses statistical data on the major industries in the Netherlands to show that companies attain high emissions-reduction percentages at decreasing costs. Finally, an attempt is made to explain the fast, technological progress.

5.1 Productivity and Environmental Demands

Several studies conducted in the 1980s addressed the negative effects of environmental policy on productivity. The stagnation of economies in the 1970s triggered research on the factors that cause productivity slow-down. This slow-down is explained by the increase in prices. Using statistical data about output and input prices, the researchers assessed which factors cause a price increase. One of the factors identified in the studies is the price increase caused by the rising costs of emissions reduction under the influence of environmental policy. The share of environmental policy in productivity slow-down is estimated in comparison with other factors, such as high oil prices and labor costs (Havenman and Christainsen, 1985; Gallop and Roberts, 1983; Barbera and McConnell, 1986). The scholars assumed that the factors could be separated, though in reality they interact with each other. In addition, they supposed that policy has no relevant positive productivity effects such as economies of scale, although scale imperfections have been found (McCain, 1978). The data has been used in the period of upcoming environmental policy and a few interlinked factors that negatively influence productivity such as the increase in oil prices and labor costs in national income, an expanding public sector and so on. The results suggest that the share of environmental policy in the national productivity slow-down is between 8% and 44% and that regulated sectors are more affected than non-regulated sectors. However, it is argued that these results are caused by the interlinked factors and that the negative effects are temporary because economies adapt gradually to the regulations (Jaffe et al., 1995).

A number of studies on the effects of environmental policy on productivity with data from one decade later have provided nuances. The negative productivity effects that have been found are very small. A study from Germany has indicated that the largest negative productivity effect of environmental policy is 2.5% in the paper industry, but the observed effects are so small that the negative effects can be counteracted by the positive side effects that have not been studied, such as changes of scale, effects on research and so on (Conrad and Wastl, 1995). A few other studies also highlighted the positive side effects. A study into the SO_2 policy in the electric power sector in the United States that examined the effects of environmental policy on capital and labor savings illustrated that the main effect of the policy was the sector's increase of scale. It is argued that the positive effects on scale can counterbalance the negative effect that higher emissions reduction costs have on prices. Another statistically observed effect is that more efforts are put into technology development because companies intensify R&D (Yaisawarng and Klein, 1994). Experts have also analyzed if there is a positive relation between R&D, patents and the stringency of environmental policy in the United States. The study has indicated that the sectors with high emissions-reduction costs spend more on R&D, but no significant effect on patents was found. It was concluded that environmental policy has an adverse short-term effect but it also provides a positive long-term side effect due to more R&D (Jaffe and Palmer, 1997). Environmental policy's positive effects on productivity have also been assessed. It has been found that stricter environmental demands invoke environmental innovations within the pulp and paper industry. Innovative companies in the sector have gained a cost advantage in comparison to companies that have used available technologies from the past. It is suggested that strict environmental demands provide an incentive for competition and that this has positive effects on productivity (Gray and Shadbegian, 1998).

The ambivalent results about the effects of environmental policy on productivity also emerge from studies on products. Environmental policy's negative effects on products are underscored by scholars who assume a trade-off between environmental R&D and R&D for products that can be sold on markets because the former crowds out the latter. Hence, using the data from the 1970s, studies based on this assumption argued that the prices of strictly regulated products such as medicines and pesticides increased faster than the prices of less regulated products (Hartje, 1984; Schultze, 1985; Eads, 1990). This view is corrected by studies that show consumers' preference for regulated products. For example, a study on R&D on safe packaging for medicines found that safer packaging for medicines increased the value of products in comparison with products packed in a regular manner. This was because consumers found that the safely packed products had a higher value than the regularly packed ones, albeit there is a limit for adding value through safer packaging. The study concluded that environmental policy contributes to added value in sales as long as it is not over-regulated (Kip Viscusi and Moore, 1993). Beforehand, there is no reason to expect any welfare losses or productivity decrease if there is a demand for environmentally improved products or loss prevention saves money.

Several studies searched for environment policy's positive effects on productivity. Many have argued that environmental technologies reduce pollution alongside the beneficial savings of energy and materials, calling for pollution prevention and cleaner technologies (Huisingh et al., 1985; Sarokin et al., 1985; Dieleman and De Hoo, 1993; Riele and Zweers, 1994; Van Berkel, 1996). It is also underlined that strict environmental policy triggers technology development and enlarges the domestic market, thus creating opportunities to gain experience with new technologies that is needed for exports (Weidner, 1985; Klaassen and Nentjes, 1986; Krozer and Nentjes, 1988). The relevance of this is important regarding the growing global market for environmental technologies particularly for the treatment of air and water emissions, renewable-energy technology and waste processing. Analyses have shown that the growth is not limited to the industrialized countries, but can also be found in developing countries. The industrialized countries are net exporters because they gain experience at home and they own most of the patents. This enables companies to develop environmental technologies and to secure a good export position for the future as environmental innovation is important in exports. Overall, the machinery and equipment industries become the main developers and exporters of environmental technologies and gain income from it. At the same time, the chemical industry, basic metal, refineries and food industries that are the largest polluters lose income (OTA, 1994; Lanjouw and Mody, 1996). It is also argued that management is triggered by environmental policy to audit processes, products and operations gaining improvement in production and putting more efforts in innovations. These, in turn, provide a competitive advantage to the innovating companies (Porter and Van der Linden, 1995).

The studies provide mixed results. Some suggest that environmental policy has limited, negative effects on productivity. The positive effects of higher R&D are also disputable when the low R&D output, measured by patents, indicates less effective R&D expenditures. Some sectors gain such as equipment manufacturing, but the polluting sectors lose, albeit the innovative companies in the polluting sectors can gain competitive advantage. In addition, the cases of the positive effects of innovative efforts are too abundant to neglect but can be overoptimistic regarding the stricter environmental policy. A key issue in the debate is whether stricter environmental policy causes cost increases in the polluting sectors or decreases due to technological progress. In the case of decreasing costs it can be argued that the policy's positive side effects on competition outweigh the costs with an overall positive effect on productivity. This is analyzed in the case of the main polluting industries in the Netherlands.

5.2 Effect-increasing and Cost-reducing Technological Progress

The starting point is that stricter environmental policy increases the cost, whereas cost-reducing technological progress can counteract this. Yet, what is the result of these forces? To answer this question, a statistical study has been done on the costs of emissions reduction in selected industries in the Netherlands during the period 1980 to 2002. The costs of stricter environmental policy without technological progress are calculated using the MOSES model (Jantzen, 1992). The cost functions of emissions reduction in the model are introduced in Chapter 3. The unit costs are calculated before the implementation of environmental technologies at sources in the industries. This means these are the expected costs of stricter environmental policy (*ex ante*). The model's unit costs are compared with the unit costs calculated with national statistics. The statistiscs measure the costs that the policy had on implementation of environmental technologies (*ex post*). The statistically observed costs include technological progress during the investment period and during the use of environmental technologies (Krozer, 2002; Krozer and Nentjes, 2007). Technological progress is defined as the difference in unit costs between the model's estimates and the statistical observation. This is shown in Figure 5.1. The figure shows the cost-reducing technological progress in relation to emissions-reduction percentage over several years, which means the annual change of the unit costs. The changes in emissions-reduction percentage during this period indicate the effect-increasing technological progress.

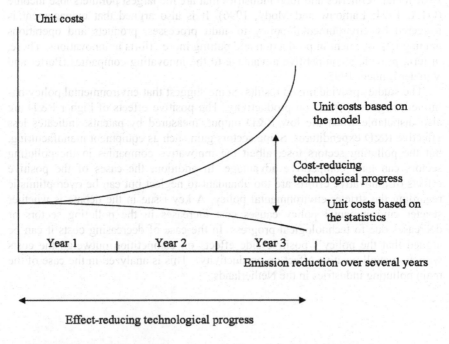

Figure 5.1. Explanation of the assessment of cost-reducing technological progress

The statistics provide a register of emissions after treatment at the sources. The available statistical data are only suitable for analyses on the acidifying emissions released to air in refineries, the chemical industry, the basic metal and electrical power sector as well as biodegradable emissions released to water by the food industry and by refineries together with chemical industries because the data do not allow separation of these two groups (CBS, *Kwartaalberichten* 1987–2005). Other emissions and industry data are unfortunately insufficient but they are also less relevant in light of the environmental policy of that time. The data do not allow an assessment of emissions reduction costs of every single pollutant but rather the impacts. The acidification impact is measured in acid equivalents (H^+), that is NO_x equals 1 H^+ and SO_2 equals 0.7 H^+. The selected four industries cover about 90% of the total industrial SO_2 emissions and about 85% of total industrial NO_x emissions in 2002. This is about 86% of the acidification impact of all industries in the Netherlands. The impact of biodegradable emissions in the selected industries, measured in the Netherlands in inhabitant equivalents (i.e.), covered about 82% of total industrial impact in 2002. For convenience, the term emissions instead of impacts is used.

Emissions reduction is the difference between emissions that would occur without treatment and the registered emissions. The untreated emissions are not registered. Therefore, the untreated emissions are calculated based on material output in physical terms on the assumption that the total physical output determines the total amount of untreated emissions. The index of material output is calculated per sector using national statistics (CBS, Statistisch Zakboek 1981–2005). The indices show that annual untreated emissions increased during the period 1980–2002; the lowest annual increase is in chemical industries by 0.014%, the highest is in electric power by 0.85% a year, but in all sectors the physical increase is far below the growth of value.[4] The physical output is chosen instead of the value because it gives a better approximation of the untreated emissions and it is less sensitive to fluctuating prices. For example, the composition of refineries' physical output in the analyzed period shows minor changes: an increase of naphtha and gasoline output from 31% to 43% of the total material output, a stable share of diesel output and a decrease of heavy fuel oil from 34% to 23% of the total. The material index equaled 100 in 1980 which means that the untreated emissions are equal to 100. It is assumed that emissions reduction started after that year, which matches the start of the environmental policy on acidification. The date is less precise in the case of biodegradable matter in wastewater because the policy had already started in 1974. The emissions reduction equals the calculated untreated emissions based on the material index minus the registered emissions found in the

---

[4] The material index in refineries is based on the sum of production of naphtha, gasoline, gas, diesel and heavy oil. In the chemical industry (during 1980–1996) it is based on the sum of ammonia and benzene, which are two basic chemicals used in many other products. Thereafter, the trend is extrapolated on the basis of the average sales value in the period 1993–1996 because the physical data are not available. In the basic metal sector, the sum of iron and steel production is used. In the electric power sector, electricity output is used. In the food industries, the sum of many food and beverage products is used.

statistics. The emission reduction percentage equals the emissions reduction divided by the untreated emissions in that year times 100.

The total costs of emissions reduction are based on national statistics (CBS, *Milieukosten van Bedrijven* 1980–1996 and later *Statline*). Some absent data are interpolated. All data are calculated in 1980 prices based on the industries' total sales. It is assumed that all air emissions' costs address the acidifying emissions and that all wastewater costs address biodegradable matter. This is an acceptable assumption because all other emissions reduction was less relevant at that time. The unit costs are calculated annually. This is done as follows: the total statistical costs in constant prices in the sector divided by the total calculated emissions reduction in the sector. The outcomes are presented for 1985–2002 because earlier emissions data are unreliable. The more recent data are not sufficiently available.

Technological progress is usually calculated by cost reduction per unit of output in time, which is based on Arrow's (1962) work on the learning function of technological progress. It is formally:

$$c(t) = c(0) \cdot e^{-kt} \tag{5.1}$$

$$k = \ln \frac{c(t)}{c(0)} \cdot t \tag{5.2}$$

where $c_{(0)}$ is the unit costs at the start; $c_{(t)}$ the unit costs in t year, k is the coefficient of the progress and t is number of years, e is the Euler number.

The output in environmental technology is emissions-reduction percentage. An additional emissions-reduction percentage entails costs, but the relation between these two is not straightforward. Hence, technological progress in environmental technology can be directed towards an increasing emissions-reduction percentage (effect-increasing technological progress) and towards reduction of unit costs (cost-reducing technological progress). These two directions are not necessarily interchangeable and therefore they are measured separately. As an analogy to Arrow's coefficient of technological progress, effect-increasing technological progress is presented by the average annual emissions-reduction percentage during the analyzed period that is given by the effect coefficient $k_{e_{it}}$. The cost-reducing technological progress is presented by the cost coefficient that is the average annual increase or decrease of the unit costs per per cent emissions reduction k_{c_e}.

The cost-reducing progress is also presented by the cost coefficient in time k_{c_t} that is the average annual unit cost increase (positive) or average unit cost decrease (negative).

The cost coefficient based on the model's data without the technological progress is compared with the cost coefficient calculated with the statistical data including the technological progress[5]. It is formally:

---

[5] A comparison between the model data and the statistical data cannot be made for the effect increasing coefficient because it is not known which technologies are implemented. In addition, the coefficient of cost-reducing progress per year can only be derived from the coefficient per per cent emissions reduction because the model data and the statistical data do not specify the time path for implementation of the environmental technologies. For the

Material growth:

$$m_{i+i} = m_i \cdot k_m{}^t \tag{5.3}$$

$$\frac{m_{i+i}}{m_i} = k_m{}^t \tag{5.4}$$

$$k_m = \frac{1}{(t-1)} \sum_{i=1}^{n} \ln\left(\frac{m_{i+1}}{m_i}\right) \tag{5.5}$$

Emissions-reduction percentage

$$q_i = A \cdot f(m_i) \tag{5.6}$$

$$d_i = \frac{q_i - l_i}{q_i} \cdot 100 \tag{5.7}$$

$$d_{t+i} = d_r \cdot ke^t \tag{5.8}$$

The effect-increasing technological progress is with coefficient:
For $t=1$

$$\ln\left(\frac{d_{i+1}}{d_i}\right) = \ln(k_e) \tag{5.9}$$

$$k_{e_r} = \frac{1}{(t-1)} \sum_{i=1}^{n} \ln\left(\frac{d_{i+1}}{d_i}\right) \tag{5.10}$$

The cost reducing technological progress with coefficient per per cent emission reduction:

$$c_i = \frac{C}{e_i} \tag{5.11}$$

$$c_i = A \cdot f(d_i) \tag{5.12}$$

$$c_i = c_{i+1} \cdot k_{c_e}{}^d \tag{5.13}$$

$$\frac{c_{i+1}}{c_i} = k_{c_e}{}^d \tag{5.14}$$

$$\ln\left(\frac{c_{i+1}}{c_i}\right) = d \cdot \ln(k_{c_e}) \tag{5.15}$$

$$k_{c_e} = \frac{1}{(d-1)} \cdot \sum_{i=1}^{n} \ln\left(\frac{c_{i+1}}{c_i}\right) \tag{5.16}$$

The cost-reducing technological progress with coefficient in time:

$$d_i = f(t_i) \tag{5.17}$$

For

$$d_a = \frac{d_{i1} + d_{i2} + d_{i3}}{3} \qquad \text{and } d_n = d_{17} \tag{5.18}$$

---

first three years average percentage emission reduction is used to reduce spread between the years.

$$D_t = d_n - d_a \tag{5.19}$$

$$k_{c_t} = \frac{k_{c_e} \cdot D_t}{t} \qquad \text{for } t = 17 \tag{5.20}$$

Half-value time:

$$c_{t+i} = c_t \cdot e^{k_c i} \tag{5.21}$$

$$C_{0.5} = \frac{c_{t+i}}{c_t} = 0.5 \tag{5.22}$$

$$C_{0.5} = \ln 0.5 = i \cdot \ln(k_{c_t}) \tag{5.23}$$

Where m material in physical terms, A is a constant, e is the emission reduction, q is the untreated emissions, l is the residual emissions, d is demanded emissions-reduction percentage, C is the total costs, c is the unit costs, n is the source technology combinations or t years, $C_{0.5}$ is the half value time.

The statistically observed emissions reduction and the costs in euros per acid equivalent and per inhabitant equivalent (i.e.) are presented in Table 5.1. The progress in emissions reduction is impressive between 1985 and 2002: from 11% to 40% annual emissions reduction at the start up to 61% and even 82% annual emissions reduction. The increase in emissions-reduction percentage is a trend in all industries albeit some minor fluctuations are found (correlations between the empirical emissions-reduction percentage and the calculated ones with k_{et} are high, R^2 lowest 0.84 in basic metal up to 0.97 in refineries). The unit costs substantially decreased in all sectors. The decrease is a trend in a few industries, notably reduction of acidifying emissions in refineries and the basic metal sector, as well as reduction of biodegradable matter in the chemical industry with refineries (correlations between the empirical unit costs and the calculated ones with k_{c_t} are high, respectively R^2: 0.72, 0.89 and 0.78). The unit costs in other industries fluctuate. Strong cost-reducing progress is found in all industries but there are also large differences between the industries with regard to additional emission reduction between 1985 and 2002 and the cost reduction.

Table 5.1. Emissions-reduction percentage and unit costs in € per H$^+$ for acidifying emissions and in € per, i.e. for biodegradable emissions, with material growth $k_m(t)$, R^2 correlations between empirical data and extrapolations based on the coefficients.

Years	Acidifying emissions released to air								Biodegradable emissions released to water			
	Refineries: $k_m(t)$ 2.2%		Chemical: $k_m(t)$ –0.2%		Basic metal: $k_m(t)$ 0.9%		Electric power: $k_m(t)$ 2.4%		Food industry: $k_m(t)$ 0.4%		Chemical with refineries: $k_m(t)$ 1.9%	
1985	11%	2.9	39%	0.9	36%	7.1	48%	0.20	40%	44	43%	244
1986	32%	1.1	37%	1.0	22%	11.2	57%	0.18	33%	51	46%	179
1987	27%	1.7	45%	0.8	26%	8.9	57%	0.27	33%	49	49%	166
1988	26%	1.8	46%	0.7	33%	6.4	58%	0.34	39%	38	55%	135
1989	28%	1.7	53%	0.7	32%	6.4	62%	0.38	44%	32	61%	120
1990	28%	1.8	62%	0.6	26%	8.5	66%	0.40	42%	37	60%	131
1991	38%	1.4	65%	0.7	22%	10.4	72%	0.35	65%	24	69%	158
1992	42%	1.2	63%	0.9	32%	6.1	73%	0.32	65%	25	69%	164
1993	41%	1.2	68%	0.9	40%	4.0	77%	0.32	66%	26	70%	158
1994	48%	1.1	81%	0.8	59%	2.3	74%	0.46	67%	28	72%	159
1995	50%	1.0	81%	0.7	61%	1.7	79%	0.43	72%	25	82%	126
1996	50%	0.8	80%	0.7	42%	2.2	80%	0.42	70%	30	82%	118
1997	52%	0.4	77%	0.3	46%	1.7	86%	0.17	73%	21	82%	59
1998	68%	0.6	79%	0.3	52%	1.4	84%	0.16	65%	24	82%	56
1999	68%	0.7	85%	0.3	59%	1.2	86%	0.31	72%	22	81%	58
2000	78%	0.6	78%	0.6	55%	1.3	83%	0.29	73%	21	81%	56
2001	78%	0.6	82%	0.6	61%	1.0	80%	0.15	71%	48	81%	54
2002	80%	0.6	84%	0.6	61%	0.8	82%	0.03	72%	24	81%	54
R^2	0.97	0.87	0.89	0.62	0.84	0.86	0.90	0.20	0.85	0.60	0.90	0.88

The coefficients of the technological progress are presented in Table 5.2. The coefficients of the effect-increasing technological progress (column 2) vary from 3.0% a year in the basic metal sector to 11.5% a year in the refinery industry. The coefficients of the cost-increasing technological progress are given in columns 3 to 6 (the projected cost coefficients for biodegradable matter without technological progress are equal because of insufficient data in the model). The cost coefficients without technological progress are the average increasing unit costs per additional emissions-reduction percentage (column 3). The cost coefficient with technological progress are the average decreasing unit costs (column 4). The total cost coefficient per emissions-reduction percentage is the sum of these items (column 5). The coefficients per year are the average reduction of the unit cost in the observed period, which is derived from the cost coefficient per per cent emission reduction (column 6). The half-value period is the number of years until the unit cost is below 50% of the initial value (column 7). For example in refineries, the model data project the average unit cost increase by 3.7% per per cent emissions reduction (column 2) implying an increase from € 2.90 per H$^+$ in 1985 to € 5.4 per

H^+ in 2002. However, the observed costs decrease to € 0.60 per H^+, which means a cost coefficient of 3.1% per per cent emissions reduction. The total cost coefficient is thus 6.8% per per cent emissions reduction. The additional emissions-reduction percentage is the difference between the average of the first three years that is 24% emission reduction and the latest 80% emission reduction; the difference is thus 56% emission reduction. This reduction is realized in 17 years, which gives the annual cost reducing coefficient of 6.8% * 56% /17 = 22.8%. The half-value period between the expected and the observed technological progress is about three years.

Cost-reducing technological progress is found in all sectors. The greatest progress in acidification is found in the refineries and basic metal sector. Unit costs are cut in half in three years. The least progress is found in the chemical industry, which reduced unit costs by 50% in 4 years. Progress in the reduction of biodegradable matter is lower and the half value time is longer. In 17 years, the emissions-reduction percentage in the industries has increased by a factor of two to four and the unit costs have been reduced annually by 8% to even 23% in comparison with projected costs. The cost-reducing progress of environmental technologies is usually well above 8% per year alongside an increase of emissions-reduction percentage of more than 3% a year. The assumption that emissions reduction causes exponentially increasing costs is not supported empirically. The results need explanation.

Table 5.2. The effect-increasing and the cost-reducing technological progress, respectively in per cent per emissions reduction and unit costs per year

Emission sources	Observed effect increase per year	Cost increase or cost reduction per per cent emissions reduction			Cost-reducing progress per year	Half-value period in years
		Projected cost increase (model)	Observed cost reduction (statistics)	Cost-reducing progress (total)		
(1)	(2)	(3)	(4)	(5)	(6)	(7)
Acidifying emissions						
Refineries	11.5%	3.7%	3.1%	6.8%	22.8%	3
Chemical industry	4.5%	6.1%	0.8%	6.9%	17.6%	4
Basic metal sector	3.0%	6.3%	5.4%	11.8%	22.7%	3
Electrical power industry	3.1%	5.0%	8%	13.0%	21.3%	4
Biodegradable matter						
Food industry	3.5%	2.3%(*)	1.6%	3.9%	8.4%	8
Chemical industry with refineries	3.7%	2.3%(*)	4.3%	6.6%	13.7%	6
(*) food and chemical with refineries equal because of imperfect model data						

5.3 Factors that Contribute to Technological Progress

How can the effect-increasing and the cost-reducing progress in environmental technology be explained? The explanation can only be tentative because there are not many empirical economic studies supporting it. The effect-increasing progress can plausibly be explained by R&D on more effective environmental technologies followed by their dissemination invoked by enforcement of stricter environmental policy. The cost-reducing progress, however, needs elaboration because the dissemination implies more investments entailing higher annual capital costs and more complex operations to comply with the demands. These, in turn, cause higher operational costs.

An explanation, based on the neoclassical theory, is that economies of scale counteract the costs because the costs increase less quickly than the scale of emissions reduction and because of the market growth due to the stricter policy that enables technology suppliers to standardize production. This has been found for the market of photovoltaic cells estimating 10% to 20% cost reduction per doubling of sales (Nakicenovic 2002). Along this line, there should be a link between the cost-reducing coefficients and the effect-increasing coefficients that indicate the scale of emissions reduction over time. Another argument, one that largely follows the evolutionary theory, underscores R&D functions in technology development (Faber and Melle, 2004). There is empirical support for this view. For example, a 15% annual cost reduction of photovoltaic cells during 1974 to 1999 in Japan is explained by know-how spillover from the developers to the manufacturers of the photovoltaics, production scale due to market growth and accumulated know-how in manufacturing industries (Watanabe et al., 2002). In this case, investment costs in environmental technologies should grow more slowly than the scale increase indicated by the effect-increasing coefficients. The third train of thought focuses on operations of environmental management at emission sources, which can be expressed by the term "learning by using" after Rosenberg (1982). There is some empirical support for this view as well. A study into desulfurization technologies in the electrical power sector in the United States has estimated 20% to 30% cost reduction in a few years after installation due to better equipment use (Wiersma, 1989, p. 230–235). Cost reduction in use is also found for other technologies, albeit with differences between the technologies: the highest, 7.5% annual cost reduction for low-NO_x burners, versus 2.5% for dephosphating at water treatment plants (Jantzen et al., 1995).[6] This argumentation implies that the total annual costs should increase more slowly than the investments.

---

[6]   In another study three main factors are hypothesized: market volume, the add-on technologies because production can be standardized contrary to process-specific integrated technologies and R&D-intensive technologies because improvements are possible [Honig et al., 2001], but evidence can be counterintuitive. For example, high-efficiency boilers, a product-integrated technology, became cheaper after a few years, but the price of catalysts for car exhaust systems, which is add-on technology with a huge market, hardly changed.

Following these three viewpoints, it is possible to estimate the factors in the technological progress in the selected industries. The following hypotheses are put forward. The importance of scale would be indicated by the link between the effect-increasing coefficients and cost-reducing coefficients across the sectors. The major role of R&D would be reflected by the lower growth of cumulative investment costs compared to the effect-increasing coefficients across the sectors. In case of learning, the annual costs would increase at a lower rate than the investments costs and match the cost-reducing coefficients across the sectors. Table 5.3 shows the cost-reducing coefficients, the effect-increasing coefficients, the growth of cumulative investments and the growth of the total costs.

Economies of scale do not explain sufficiently the cost-reducing technological progress because there is no cross-sector link between growth of scale and unit-cost reduction, which confirms the findings based on cost functions of emissions reduction presented in Chapter 3. The effect-increasing coefficient is larger in some sectors and smaller in others than the cost coefficient. (The correlation between these is low, $R^2 = 0.38$). Even less clear is the role of R&D in the cost-reducing progress. Cumulative investments increase faster than the cost-reducing coefficient in all sectors. (The correlation between these is low, $R^2 = 0.004$). A major factor is apparently the learning process at the emission sources regarding the larger increase of the investment costs alongside minor total annual cost increases and in some sectors, even a decrease in the total annual costs. (The correlation between these is, $R^2 = 0.47$). The findings suggest that economies of scale and improvements in environmental management are among the main factors in the cost-reducing technological progress.

Table 5.3. Factors to explain the cost-reducing technological progress in selected industries

Factors	Acidifying emissions				Biodegradable matter	
	Refineries	Chemical industry	Basic metal sector	Electrical power industry	Food industry	Chemical & refineries
Cost-reduction coefficient	−22.8%	−17.8%	−22.7%	−21.3%	−8.4%	−13.7%
Effect-increasing coefficient	11.5%	4.5%	3.0%	3.1%	3.5%	3.7%
Growth cumulative investment	18.6%	18.7%	12.6%	20.3%	16.9%	16.6%
Growth total costs	4.7%	2.0%	−8.2%	−6.0%	0.3%	−3.0%

An explanation of this rather odd outcome could be that polluting companies anticipate the stricter regulations with investments that "overshoot" the expected policy demand. They take on (voluntarily or by obligation) larger efforts than those that are strictly needed to comply with the policy, but this provides a license for operations minimizing laborious negotiations with policy makers and enables production growth. Once the companies have received the license, the environmental management finds ample opportunities to reduce costs by tuning the installed technologies to the expanding production and through better operations with the technologies without risking non-compliance with the license. This explanation is supported by the observation of statistical data in a few sectors that a strong increase in investments corresponds with a strong decrease in the unit costs with a time lag.

5.4 Conclusion

The question of whether the industries that face environmental policies are risking high costs and a competitive disadvantage or if competitive industries are able to respond effectively and efficiently is largely answered in favor of the latter, though economic benefits of environmental technology are not found because they are statistically not measured. During the period 1985 to 2002, the polluting industries in the Netherlands increased the emissions-reduction percentage by a factor of two to four; which approaches the demand for pollution reduction that does not preclude sustainable development. This far-reaching emissions reduction has been achieved through huge investments in environmental technologies. The investments, however, did not cause much cost increase. The total costs have stabilized and even diminished in some sectors. The costs per unit emissions reduction are reduced by 8 to 23 per cent annually in comparison with the projection of environmental policy. The intuitive argument that polluting companies first use cheap technologies when confronted with the policy, followed by more expensive ones is not supported by statistical evidence. It seems that the industries voluntarily or by obligation install the technologies that license production growth alongside emissions reduction and they find ample opportunities to reduce the costs during use of environmental technologies. The reality is more complex than the "common wisdom".

Stricter environmental policy invokes investments to reduce emissions that increase capital costs, but it does not imply higher total annual costs nor growing costs per unit emissions reduction as is presented in handbooks on environmental economics. The analyzed industries have found ways to invest in new, more efficient environmental technologies and go beyond that by process changes to reduce the costs. The industries' different levels of technological progress are difficult to explain. Research on environmental innovations is still in its infancy. Plenty of theoretical propositions can be found but are seldom tested empirically, whilst scarce empirical studies are usually only case studies. Many questions remain. It is unclear why certain sectors and companies can reach a high rate of environmental innovations while others cannot attain it, what factors and policies invoke fast technological progress, how industries can benefit from environmental innovations and so on. The key conclusion remains that stricter environmental policies collide in the short term with companies' profitability but companies do adapt well to changes in cost factors invoked by the demands if they have the freedom to select the solutions.

6

Policy Demands and Environmental Management

Emissions reduction at steadily lower unit costs due to environmental innovations, shown in Chapter 5 poses a question: What conditions enable companies to use innovations that reduce the costs of emissions reduction, or even provide a benefit? It is argued that not only strict demands create favorable conditions for innovations, but rather that capable environmental management in companies is also needed. Company case studies suggest that such conditions are not unusual.

6.1 Environmental Management

It is widely observed that companies usually accept higher emissions-reduction costs than the lowest cost attainable with innovative solutions. It is usually explained by bounded rationality in environmental management, which is discussed in Chapter 2. The explanation is that decision makers aim to satisfy their aspiration through plausibly available solutions rather than maximize profitability by searching for all possible solutions. In the case of environmental managers, the satisfying decision is to comply with policy demands even at a high cost as long as it does not affect the company's profitability. Regarding decisions about the use of technologies, the satisfying decision usually entails use of available technologies from the past, which are BAT technologies prepared by policy makers before setting and enforcing the demands. Such a decision would be made, despite the chance that some lower-cost innovative solutions could be found, as long as profitability is not endangered and managers can give its attention to other issues.

Several studies underpin this view. A study on cleaner technologies has shown that environmental managers usually make decisions based on risk perceptions rather than on a costs-benefits assessment. This means that potentially rewarding, cleaner technologies are often not used as they are perceived as risky (Kemp et al., 1992). Something similar is found in the energy sector in developing countries. Calculations show that 15% to 20% of energy use during energy production can be saved. This energy benefit is not utilized because an increase in production volume is prioritized above lower costs (Cavendish and Anderson, 1994). A study using U.S. statistical data on effects of energy prices on substitution of capital goods has also indicated that the dissemination of energy-saving technology is slower than should be expected on the basis of past prices. This is explained by management's expectation that future energy prices will eventually drop (Thomson and Taylor 1995). Another study on investment decisions in energy saving using EU statistical data has shown that the net beneficial technologies are systematically unused because management is preoccupied with cost factors other than energy, such as labor. It also suggests they go unused because management has little experience with energy-saving technologies (Velthuijsen, 1995). The importance of such a management preoccupation is confirmed by analysis of companies' voluntary participation in energy-saving programs regarding the participation that reflected regulatory intensity rather than energy intensity. This analysis has concluded that companies' attention is the major factor influencing energy saving (DeCanio and Watkins, 1998). The general observations are that the managers tend to avoid the risks connected with the use of innovations and that they are focused on factors other than environmental performance. As a result, low-risk solutions are adopted even if less-costly ones are available.

Business leaders generally support the proposition that companies can reduce emissions at lower cost through more effective energy and resource use (higher ecoefficiency). This would happen if management were not confronted with unreasonable demands and would have the freedom to adjust environmental management to specific business practice (Schmidheiny, 1992; DeSimone and Popoff, 1997; Voges and Veerkamp, 2002). Along these lines, performance standards are introduced such as ISO. A capable management can sort out the optimal solutions that reduce emissions in an efficient way and the policies, in this view, should refrain from direct interventions but could strengthen managerial capabilities by consultancies to companies and the demonstration of environmental innovations because these reduce the risks connected with innovations (Rennings et al., 2006; Jorna, 2006). An assumption of this view is that emissions reduction can be beneficial if sound solutions are used.

Another view emerges from studies on the effects of policy making on environmental innovations. These studies show a disparity between economic and environmental objectives, which can only be merged under exceptional situations, like reuse of an emission as a high-price resource. Environmental innovations are rarely found to be profitable or to be spontaneously implemented. Therefore, the studies advocate enforcement of far-reaching environmental demands to invoke use of the innovations alongside pricing of emissions and free choice in technology implementation, which, in some cases, can even generate a benefit due to the resource savings. The undemanding policy, on the contrary, can create a self-fulfilling prophecy of costly and risky environmental policy because being innovative is not rewarding (Saviotti, 2005; Cerin, 2006; Vollenbergh, 2007).

The issue, then, is how to understand the bias of environmental managers regarding possibilities of innovations. The question is whether strict environmental demands invoke use of lower-cost environmental innovations as policy analysts suggest, or if it is rather the innovative behavior of environmental managers that brings low-cost solutions.

6.2 Compliance Versus Anticipation

When a company is confronted with an emerging environmental demand during policy preparation, the management can opt to comply with the demand using an available technology from the past that is usually a BAT suggested by the policy maker. Another option is to anticipate enforcement of the demands by searching for an innovative option, which is uncertain in terms of costs and benefits but possibly rewarding in comparison with the available technology (Hartje and Lurie, 1984; Zimmerman, 1985; Driel and Krozer, 1987). The choice between compliance by the available technologies and anticipation strategies by the innovations is used to elaborate conditions that favor decisions for environmental innovations.

If environmental managers opt for innovation, they must search among the options, select the most attractive one, test its practicality in use and learn how to operate the new technology. All of these efforts are needed before the actual investment that entails installation of the new equipment. Hence, the option of using an innovation has the risk of costs connected with all the efforts related to preparation of the investment before installation and use of the environmental technology. Such costs, called the costs of change-over, are needed to shift from the best available technologies from the past to an innovation. The costs of change-over must be made despite uncertainty regarding the innovation's performance compared to the available technology during use. The high costs of change-over and the uncertain benefits of the innovations during use explain the bias against environmental innovations. This is called bounded rationality but actually it can be the best choice given the costs of change-over and uncertainties about the innovations.

The bias can be invoked by the regulations. As policy makers strengthen the preparation of available technologies to assure compliance with an environmental demand, it becomes less attractive for environmental managers to raise the cost of change-over and use environmental innovations. By setting environmental demands far away in the future or through imaginary technologies the policy makers create uncertainty. As a consequence of the discounting that reflects uncertainty, such environmental demands, even the very strict ones, discourage use of innovation. This is because the present value of the savings due to the innovation compared to the available technology, becomes low and even negligible for the risky innovations because of the high discount rate.

Management deliberation regarding future policy demands can also be biased. In compliance, the environmental manager can wait until the demands are enforced and use available technologies when negotiating about a license. In anticipation, the environmental manager must raise awareness and money in the company to cover the costs of change-over before enforcement. Afterwards it has to wait and see whether the strict demands are going to be implemented and innovations save costs. The environmental management must also be capable of carrying the efforts, which is not always possible. The costs of change-over can be too high for companies with limited resources and low access to information. An economic slump limits the possibilities to raise money for the change-over and the innovations can be less rewarding. In the same way, an economic upswing makes environmental innovations more attractive because they become more rewarding due to energy and material savings and give better product quality alongside greater emissions reduction. However, the rewards are usually surpassed by high expenditures in other areas of environmental management with overall costs in the companies' accounts. It does not mean that the specific environmental innovation is insufficiently beneficial or that a specific pollution prevention effort does not pay. Instead, it suggests that overall environmental management is a cost. Examples of the deliberation about process and product changes are shown in the boxes. Similar deliberations are encountered with regard to the treatment technologies. For example, a mechanical wastewater treatment can comply with moderate water-quality demands but for compliance with stricter demands upgrading to biological treatment is needed. This makes it much more costly than the mechanical plant, which is in turn more costly than constructing the biological wastewater treatment plant right from the beginning.

An example in shipping A shipping company can comply with policy demands by reducing speed and using better route planning that saves fuel. They can also use BAT technologies to reduce NO_x. An innovative option is to redesign ships for high efficiency engines that use low-sulfur diesel or gasoil instead of high-sulfur bunker oil. This cuts fuel use and costs. This option requires high investments and skills in construction and operations. It can only be implemented through an integration of technologies when constructing new ships. The change-over to high-quality fuel is rewarding for some shipping companies (Doelman et al., 1994; Krozer and Maas, 2003).	An example in printing Printing companies can comply with demands on volatile organic compounds (VOC) from ink by BAT technologies, which include collection, regeneration and sales of VOC as a solvent to other firms. An innovative option is solvent-free ink. The use of solvent-free ink in printing calls for adaptations in the process and high investments. The advantages are savings on energy use and higher sales as some customers appreciate prints made with solvent-free inks and will pay a premium for them because they reduce health hazards involved in printing. The premium prices can outweigh the change-over costs. (Kothuis et al., 1997; Hofman, 2001).

The argumentation is translated into a model for companies' environmental strategies (Krozer, 1992). The model describes the costs of environmental management facing future policy demands. The costs of environmental management that uses the available technologies increase exponentially because extra technologies must be implemented to comply with the demands. As a result, the anticipation strategy with innovations is cheaper than the compliance strategy with available technologies at strict demands, albeit less attractive in the case of uncertain demands. However, the available technologies are cheaper than innovations regarding moderate, future demands. This is because the costs of change-over are higher than the savings due to innovations, or the costs change-over are difficult to bear. The model is presented graphically in Figure 6.1.

Marginal costs of emissions reduction

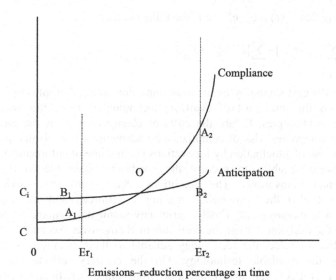

Figure 6.1. Model for environmental strategies of companies: deliberation between the compliance and anticipation strategies

The demanded emissions-reduction percentage, Er, is shown horizontally. The expected marginal costs of emissions reduction, Cr, are shown vertically. A stricter demand for emissions-reduction percentage is shown by a shift from the position Er_1 to the position Er_2 on the horizontal axis. The costs as functions of emissions reduction are shown by the lines A and B. The compliance cost function, A, is based on available technologies from the past. These costs increase exponentially. The compliance cost function can be defined by using databases on available technologies, as described in Chapter 3 eventually with adaptations for specific emission sources. The anticipation cost function, B, shows that the costs of change-over must be made before an investment at Er_2 but the costs of emissions reduction do not increase much and can even decrease in the case of beneficial activities such as energy and material savings or increased sales due to better products. The marginal cost of the anticipation strategy compared with the marginal cost of the compliance strategy is higher at the low demand, Er_1, but lower at the strict demand, Er_2. The present value of the total costs of change-over given in the figure is represented by the area A_1OB_1. The present value of the total benefits of the innovation in comparison with the available technologies is represented by the area A_2OB_2 (representing the surplus of innovation rents) The anticipation is attractive if $A_2OB_2 - A_1OB_1 > 0$. The model implies that the anticipation strategy through using innovations is usually costly with moderate demands but it can be attractive with strict environmental demands.

The deliberation of environmental managers regarding the expected surplus of innovation rents can be formalized in the following manner:

$$C_a = c_a \cdot e_a \text{ and } C_i = c_i \cdot e_i \tag{6.1}$$

Knowing that $c_i(t) = c_a \cdot e^{k_c \cdot t}$; e is the Euler number

$$S_e = \sum_{i=1}^{n}[C_a \cdot r_a^{-t}] - \sum_{i=1}^{n}[C_i \cdot r_i^{-t} + C_{c_i}] \tag{6.2}$$

for $r > 1$

where S_e is the cost saving by innovations (emission sources' surplus of innovation rents), C_a is the total cost of available technologies, C_i is the total cost of innovative technologies, C_c are the costs of change-over, r_a is the interest that reflects the companies' risk of compliance by adaptations and r_i is the interest that reflects the risk of anticipation by innovations t is the time of enforcement in years.

This model enables to specify the conditions favorable to the use of environmental innovations. The conditions address not only environmental demands but also the way policy making is enforced and the capacity of environmental management. Firstly, strict environmental demand is needed to encourage the decision to raise the costs due to change-over. Secondly, companies must be free to select the technology suitable to their situation, which is not necessarily the available technology. On the contrary, efforts put into the preparation of BAT can divert attention from more beneficial environmental innovations. Thirdly, companies can only introduce and use innovations if they can anticipate policy making. This implies that demands must be announced in advance of enforcement to allow time-consuming change-over and that the policy must be firm to reduce uncertainty. The latter implies that the regulation with imaginary solutions can delay the innovations because non-compliance can be justified. Fourthly, managers must sense urgency in order to decide in favor of the anticipation strategy. This means that demands set too far ahead of enforcement can be considered irrelevant by the management and, in consequence, the present value of savings due to innovations fades away in time, thus discouraging the change-over. Fifthly, management must be able to raise the needed funds to cover the costs of change-over. This is difficult during an economic slump and in companies that experience low profitability. Sixthly, management must be able and willing to take a risk regarding the uncertain performance of innovations during use, which depends on the company's leadership capabilities. The rewards of the anticipation through innovations can be cost saving or net benefit, even though companies bear the cost of change-over.

6.3 Benefit of Environmental Management

The model has been used in support of environmental management facing environmental policy demands. Scenarios are drafted under various assumptions about policy making and management performance and used to support companies' decisions. Three examples of scenarios are presented: on energy saving, on waste management and on technology development. All of the studies have been commissioned by companies and executed within the framework of the Institute for Applied Environmental Economics (TME). The first study, executed by Van Duijse and Krozer in 1994, indicates that integration of energy saving with treatment of emissions provides beneficial solutions for acidifying emissions and greenhouse gases. The second example, elaborated by Doelman and Krozer in 1996, illustrates that innovations in the recycling of electronic products reduce the costs of collection and recycling and that strict demands in combination with economic instruments in environmental policy facilitate beneficial innovations in recycling. In this case, the instrument is the take back of products after use with a fund. The environmental-oriented product development (ecodesign) in combination with such a policy instrument provides net benefits. The third case, done by Krozer and Lavrano in 1994, addresses technology development, which is process innovation for wood preservation. The innovation aims to produce high-grade wood without the use of chemical additives. The innovator anticipated demands for low-toxic products, reduction of CO_2 emissions and limitations in the use of high-grade wood from tropical forests. In all cases, the costs and benefits are discounted at a 110% interest (10% interest rate).

Anticipating Acidification Regulations
The question raised by a large, international company has been whether the anticipation of strict demands for SO_2 and NO_x emissions reduction through energy saving can be advantageous in comparison with emissions-reduction technology alone. The anticipation was needed because energy saving requires large investments and the company had to search and test various energy options for many subsidiaries around the world. The expected environmental demand after ten years was a 40% reduction in NO_x and 70% reduction in SO_2 in comparison with the initial situation. The initial energy use was 65 PetaJoule (PJ), 70% electricity and 30% fuel (out of 65% oil and 35% gas). The environmental demands are based on the Dutch policy (Ministerie VROM, 1988–1989, pp. 36 voor NO_x, pp. 59 for SO_2) that were less strict than the policy in Germany or Japan but more stringent than that in many other countries, including the United States.

The emissions of NO_x and SO_2 are caused by fuel combustion, particularly in boilers. The emissions can be reduced by treatment technologies. Another option is process-integrated energy saving that partly reduces the emissions and calls for fewer treatment technologies. Some energy-saving technologies are net profitable due to energy saving, but "extra" technologies cause net costs, though they can reduce emissions at lower costs than the costs of treatment. The "extra" technologies are mainly heat pumps and heat storage. They reduce fuel use but need electricity. The costs of electricity are included, but not the emissions at electrical power plants. It is found that the maximum use of energy-saving technologies is insufficient to attain the goals for NO_x and SO_2 emissions reduction. So some treatment technologies are needed in addition to energy-saving technologies, but fewer than without energy saving.

The estimated annual costs of the compliance strategy are based on the treatment of pollutants alone. The costs of the anticipation strategy cover both energy saving and treatment. The costs that are presented in Graph 6.1 are the sum of annual capital costs and operational costs after subtraction of the benefits gained due to energy-saving. The negative costs indicate the net benefit due to energy saving. The environmental demands can be attained through the combined energy saving and treatment technologies at lower costs than treatment alone. The benefits of energy saving are estimated conservatively. Internal company accounts have been much more optimistic about the benefits because a lower depreciation rate on the investments is used. An additional advantage of energy saving has been anticipation of the demands for CO_2 reduction, which was an emerging but controversial issue at the time of this research study. The company had implemented many energy-saving technologies. Hence, it attained higher ecoefficiency in reduction of acidifying emissions than solely through emissions reduction technologies. The "extra" energy-saving technologies have been postponed mainly because of uncertainties about the enforcement of environmental demands in many countries. This has impeded dissemination of the new energy-saving technologies.

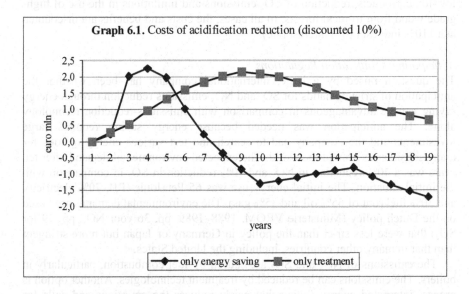

Graph 6.1. Costs of acidification reduction (discounted 10%)

Anticipating Recycling Regulations

The benefits of product redesign in combination with economic instruments in policy making are presented by a case study on recycling of post-consumer electronic products. In this case, the costs of disposal and recycling of electronic products in the European Union (EU) are assessed. The study was requested by a consortium of electronic companies to anticipate stricter EU disposal regulations. The study followed an announcement made by a few EU countries prohibiting the disposal of untreated electronic waste and strict recycling demands in Germany and Sweden. The disposal method has mainly been shredding, pressing and landfill with a heavy environmental burden caused by the release of hazardous compounds. A few cost scenarios were done for the period 1992 to 2010, including logistics, recycling and sales of secondary products. The assessment was done for televisions, audiovisual equipment and computers, including the expected growth of sales. These products cover about 80% of the total market volume of electronic goods on a weight basis. Innovations to enhance productivity in recycling are products designed for dismantling and incorporating an identification unit in equipment to support take-back of the products, logistics and recycling. The identification unit can be a bar code with information about the product type, hazardous compounds and their location in the equipment or a more complex unit to monitor malfunctions and repairs occurring in the equipment's life cycle. The study also assessed what economic instruments could support the introduction of innovations. The costs are shown in Graph 6.2.

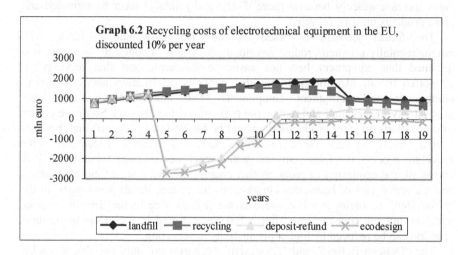

Graph 6.2 Recycling costs of electrotechnical equipment in the EU, discounted 10% per year

In the scenario "Landfill", the costs of shredding, pressing and landfill are assessed. The costs increase because more and more plastics are used that are difficult to separate and that have low value. In addition, landfill standards have become stricter, so landfill of residues becomes more expensive as well. The costs of transport and the rising costs of incineration and landfills are also considered. The average costs of landfill in 2010 in the EU are based on the average costs in Germany in 1992, at that time, the highest in Europe. In this scenario, the costs increased from € 750 million in 1992 to roughly € 2.8 billion in 2000 and even € 5 billion in 2010. The costs increase due to higher sales (particularly computers) and rising costs of landfill.

In the scenario "Recycling", the costs of collection and recycling are assessed, including the sale of materials and landfill of residues. The costs are assessed per product because of large differences between equipment. For example, the recycling of televisions is much more expensive than that of computers. The costs are based on commercial experiences in the Netherlands and compared with similar activities in Germany and Sweden (the costs in these countries are similar). In this scenario, while the costs increase as well, they are lower than in the Landfill scenario. Initially, recycling is more costly than landfill but it becomes cheaper after a few years because of increasing landfill costs.

In the "Deposit-Refund" scenario, it is assumed that specialized dealers or retailers handle collection. In addition to the recycling scenario, this case includes a deposit or a price addition for collection and recycling that is refunded when the product is returned. It is assumed that the deposit or price addition covers the costs of handling at the retail level and the recycling costs. The suggested deposit or price addition is so low (€ 9 per piece) that it does not affect sales and is implemented throughout the EU to prevent negative effects on competition. The assumed average lifetime from procurement to discharge of a television is seven years, audiovisual equipment lasts for five years and computers only three years. In the deposit refund scenario, the companies generate net income after five years because they cash in the deposit or price addition and a rent. After 10 years, the costs increase quickly because more discharged products must be refunded and more products must be recycled.

The fourth scenario is based on the "Deposit-Refund" scenario with environmentally oriented product development. In this "Ecodesign" scenario, it is assumed that equipment becomes easier to dismantle and that recycling is supported by an identification unit that contains information about hazardous compounds and equipment components. The innovations increase labor productivity in recycling and reduce the costs of residue disposal. Since the effects of innovations on labor productivity cannot be forecast, it is assumed that the costs decrease by 10% a year. There are various methods to increase productivity in the recycling process (such as semi-automated machines or even robots), but it is above all, the ecodesign of products that enables the fast dismantling of products and the prevention of hazardous compounds that reduce the disposal costs. In the "Ecodesign" scenario, initially, there is net income due to the deposit or price addition, but the cost increase after a few years found in the "Deposit-Refund" scenario can be reduced by higher productivity in recycling.

The "Deposit-Refund" and "Ecodesign" scenarios are only possible if market interests co-operate. In this way, cost reduction can be attained in comparison with landfill or recycling alone. The Ecodesign can even provide net income. There are initial costs, but advantages come later due to a combination of financial instruments with innovations through Ecodesign, sensors and recycling. In fact, co-operation between several companies in a number of EU countries have led to the introduction of a fee on electronic waste for collection and processing, Ecodesign and identification units with similar results as the projected results.

Anticipating Water Pollution Regulations

The case of technology development addresses the innovation in wood preservation called Plato (Providing Lasting Timber Option). In this case study, the costs of an innovation are compared to the costs of available technologies for wood preservation. It is proposed to upgrade the low-quality wood to high-grade wood qualities. The Plato process is based on physical manipulation of wood using pressure and heat. It was developed as a side line of R&D on chemical properties of woods as a renewable source for chemicals. Plato competes with applications of high-grade wood from tropical forests and with chemically preserved wood with toxic chemicals.

It is assessed whether Plato wood customers must pay a higher price than customers using chemical alternatives for wood preservation or customers importing tropical wood, taking into account strict environmental demands in the future. The costs of Plato wood are compared to other preservation technologies: imported high-grade (tropical) wood, chemical preservation by creosote oil, chemical preservation by copper-arsenic salts with vacuum-drying technology and copper-arsenic chemical preservation with the dipping of wood in a hot bath, lacquering wood. The costs of Plato include development, production and the use of wood but exclude the research costs as these are low in comparison with full-scale production. A distinction is made between the costs with moderate environmental demands and the costs with strict demands. With respect to tropical wood, it is assumed that wood must be harvested in a sustainable manner, requiring additional space and labor in comparison to presently intensive cutting down of trees. With respect to emissions, it is assumed that treatment technologies are applied in production to reduce the toxic emissions that are mainly PAHs and heavy metals, whereas in use, dispersion of the emissions into water and soil must be prevented. The costs are assessed including and excluding the demands for CO_2 emissions reduction. This policy is relevant, because Plato wood is condensed, so more wood must be used per unit of Plato wood produced. Before combustion, wood stores CO_2. The CO_2 stored in wood can be subtracted from CO_2 emissions in production and use. In this way the costs of CO_2 emissions reduction can be lowered.

The production of Plato wood is more expensive than chemical wood preservation but it is cheaper than importing high-grade wood. Plato wood becomes cheaper than the chemical technologies if environmental demands on toxic emissions are enforced, because the costs of pollution control of Plato wood are much lower. Lacquering is also more expensive than Plato but it is done for various reasons other than preservation. With moderate environmental demands, the costs of Plato wood are higher than the costs of available chemical technologies, but the situation changes with strict demands. The Plato process is an innovation that prevents high costs connected with strict demands. If the saving of CO_2 storage in wood is included, then Plato technology is economical because the amount of stored CO_2 in Plato wood (after subtraction of energy use in production) is larger than that offered by its alternatives. The impediments to the introduction of the Plato process are imperfections in processing that cause unstable output quality and uncertainty about the enforcement of environmental demands on toxic emissions from chemical preservation.

6.4 Conclusion

The question of whether environmental management of companies can reduce the costs connected with far-reaching environmental demands and even generate a net benefit is positively answered, albeit conditionally. The costs of the best available technology that have been prepared by policy makers, though low with moderate demands, increase quickly with stricter demands, whereas environmental innovations are less vulnerable to stricter demands and the costs hardly increase or even decrease due to energy and material saving as well as better products. Therefore, management can deliberate whether to use available technologies or bear the costs of change-over for preparation of the investment in innovation in the short run and benefit from the lower costs or a net benefit in the long run. The case studies suggest that companies bear the costs of change-over but they can attain a benefit by using the innovations, whilst the fail-soft policy discourages the use of innovations.

The costs of environmental management are available in statistics, but the benefits are absent because these are side effects of the investments in emissions reduction and because the benefits in some areas of environmental management are counteracted by the costs in other areas and finally incorporated in the companies' overall results. The benefits due to environmental innovations, therefore, must be estimated by comparing the companies' strategies with and without the innovations. The findings indicate possibilities for beneficial environmental innovations albeit under several conditions. Firstly, managers must face strict and assured policy demands, whereas the fail-soft policies obscure the need for innovations. Secondly, management must have the freedom to choose the most suitable technology, whereas by enforcing the available technology the policy obstructs innovations causing inefficiencies. Thirdly, policy demands must be announced several years before enforcement begins to enable change-over and case studies show that uncertain enforcement tempers change-over. Fourthly, companies must sense urgency to innovate, so demands set in the distant future blur the urgency. Fifthly, environmental managers must be able to raise funds for the change-over, such as a fund for the management of electronic waste based on a product fee. Sixthly, management must be able to overcome the uncertainties connected with environmental innovations. The widespread notion that enforcement of strict demands causes rapidly increasing costs in emissions reduction that cannot be avoided, should be dismissed. A trade-off between environmental and economic objectives should be acknowledged but innovative environmental managers can implement technologies that temper the cost increase caused by stricter demands, even reduce inefficiencies and generate benefits by saving resources and sales of new products alongside emissions reduction in some cases.

7

Social Demands and Environmental Management[7]

Companies face social demands for products and services related to public concerns such as environmental qualities. The question is how they can respond to such concerns regarding possible changes in product life cycles and what cost advantages can be expected in the life cycle management in comparison with the focus on manufacturing. A product's life cycle can be described as a chain of physical changes involving inputs and outputs in the production of raw materials, manufacturing, distribution, use and disposal. These entail emissions and waste at every stage.

7.1 Social Demand for Products

In the conventional view, market organization enables private demanders with subjective, individual preferences to select supplied products and to buy the one they prefer at a price that maximizes their private, individual utility ("highest value for money"). The price is the main organizing mechanism and it is a signal to suppliers about the preferences held by private demanders. The issue is how to address social demands that lack such an organizing mechanism. Social demands address common goods such as use of space, ambiance qualities, amenities, emancipation, fairness and so on. The demands are group specific with respect to work and income, age and schooling, geography, culture and so on, and they chainge over time (D'Iribarne, 1983, Lombardini, 1989).

Social demands that refer to the environmental quality of products and services address various issues such as the renewability of materials, pollution in production and use, local sourcing of resources, health impacts and so on. The economic value

---

[7] Parts of this chapter were published in Krozer, Y., (2008), *Life Cycle Costing for Innovations in Product Chains*, Journal of Cleaner Production, 16: 310–321. I am grateful to Elsevier Publishers for permission to use the material.

of these demands is notable and growing worldwide, particularly in the EU. The potential economic value is substantial because many consumers wish to buy certified, upgraded products and the number is growing. For example, the percentage of German consumers who want a label for products' environmental effects increased from 19% in 1989 to 46% in 1998 (Imkamp, 2000). Actual sales are lower than potential ones based on interests mentioned in inquiries. An estimate of product sales that aim to comply with social demands for environmental qualities is around 2% to 3% of all consumer expenditures in the EU (Hoevenagel et al., 1996; Heiskanen and Pantzar, 1997; Goedkoop et al., 1999). It represents a market value that approximates the value of environmental technologies and services. This means that minor changes in environmental demands have a large effect on environmental innovations. Coverage of the labeled products widens, like low hazards, energy saving and so on. There is an increasing number of labeled foods, toiletries and fabrics, recycled paper, low-energy electric, household equipment and transport means and so on. Furthermore, one finds sales of products that can be regarded as environmental technologies for the consumer market, for example, safe paints, water filters and fluegas treatment. The annual growth of sales among environmentally labeled products is a two-digit number in some sectors such as foods and tourism and the sales of environmental technologies for the consumer market grow fast, particularly in renewable energy (Krozer, 2002). The social demands address various steps in the life cycle of products, for example the production of raw materials (e.g. pesticide-free supply), manufacturing (e.g. prevention of hazardous compounds), retail (e.g. local products to reduce transport), use (e.g. possibilities for energy saving) and disposal (e.g. reuse of materials). Life-cycle management aims to accommodate such demands in companies' decisions, which is an application of life-cycle thinking in management (Jensen and Remmen, 2004).

The difficulty involved in anticipation of social demands compared to private demands is that product functionality, essentially the product's usefulness, not necessarily matches the social demand. Hence, the task of the designer aiming at improving functionality and the task of the life-cycle manager aiming to accommodate social demands in products, often do not go hand in hand. Some social demands match functionalities. For example, it can be argued that social demands for the pesticide-free foods comply with additional functionalities of private demands for fewer health impacts. However, it is more common that private and social demands have little to do with each other because the functionalities are similar but compliance with the social demands is costly. For example, tropical wood from "regular" and "sustainable" plantations have the same functionality but the latter is at least twice as expensive. One often finds a trade-off between private and social demands, for example, cloth preservation versus durability (Krozer, 2004). A company, therefore, must serve two markets, each one with a set of prices and qualities that can stand in opposition to each other.

Secondly, management must deal with complex products' life cycles. For example, a simple product such as a steel spoon needs a minimum of three suppliers in a row. These include the steel producer, the steel roller and the galvanizer, in parallel, packaging suppliers, followed by those involved in distribution and retail sales, use and finally disposal using recycling, incineration and landfill. Most product chains entail many more steps. Each one has many inputs. Transactions are extensive because the economic value and the risk per step in the chain are unevenly distributed. Since every unit in the chain adds to the cost, which is ultimately covered by consumers that call for lower prices with better quality, firms operating close to consumers accrue the major portion of the benefits connected with complying with social demands. They pass on the costs and risks of non-compliance to suppliers. This obviously extends the transactions. Trade and cultural barriers, vested interests that are unwilling to make changes and other organizational impediments add to the laborious negotiations.

7.2 Life-cycle Valuation

Due to growing social demands on the one hand and difficulties in resolving issues in the chain, many efforts are made to support decision making in the life cycle. In environmental science, the life-cycle assessment (LCA) is often used. Briefly, LCA covers an inventory of the material streams that are inputs (resources) and outputs (emissions), followed by impact assessments based on indices (so-called classification). Several indices are used. The index can be a ratio of a current state with a demanded one (distance to target approach, such as weighted percentage of various emissions), or based on a reference value (reference approach, for example CO_2 emissions as a reference for greenhouse gases). The indices' focus differs. Some address inputs, such as fuel use or material use, because material displacement is expected to be critical for environmental qualities. Others address an area of land on the assumption that space limits the inflow of solar energy, which is crucial for all life. Many make indices of emissions. Some indices address only a few indicators, while others cover hundreds of emissions. In addition, the loss of energy quality, called entropy, is advocated using the argument that every activity dissipates energy that can only be counteracted by an inflow of solar energy. Finally, some link environmental qualities in physical terms with socio-economic indicators. These include methods ranging from a simple ranking based on experts' and stakeholders' opinions to sophisticated methods such as the probability that public opinion influences managerial decisions (Earl, 1999) and multi-criteria analyses by hundreds of experts (Jesinghaus, 2000).

This development is not only a playground for scientists, but is meant to support decision makers in particular companies' environmental managers, though it should be noted that all methods are normative judgments and study conclusions should be treated with caution. The results are useful to find major issues in the products' chain regarding social demands, whereas comparing products between each other is risky because of uncertain data and imperfect methodologies (Krozer and Vis, 1998). Following these environmental approaches in the life cycle assessments, many efforts are put to establish methodologies that approximate the market prices. This is done on the assumption that such proxies for the market prices create an organisating mechanism for the negotiations about the social demands for environmental qualities, although not as perfect as the real market prices. Scheme 7.1 illustrates the variety of methods in the environmental and in the economic approaches.

Scheme 7.1. Typology of life-cycle assessment methods (with an example)

	Input-oriented	Output-oriented	Social-effect-oriented
"Comparative"	MIPS	CML factors	"Produktlinien"
"Norm-based"	Energy benchmarks	"Öko-punkte"	Costs of demands

Environmental-economic assessment methods of social demands based on social-effect-oriented approached			
Direct cost	Life-cycle costing (also total cost of ownership/total cost assessment)		
Indirect cost	Contingent valuation	Total cost assessment	Control costs

The interests to link the environmental with economic issues invoked various LCA-related economic decision-support methods. A comprehensive review done in the mid-1990s showed more than 50 methods (Tuulenheimo et al., 1996). These are presented under names such as total cost assessment, full cost accounting, and total costs of ownership. We call them life-cycle costing (LCC) because it is now common in managerial literature (Dhillon, 1989; Fabrycki and Blanchard, 1991; Drury, 2001). The LCC is used in companies' decision making on major investments and for the life cycle of products as a tool in life-cycle management. LCC covers assessments of the costs in all steps of the life cycle during product lifetimes. It covers the definition of the main cost factors (called "cost drivers") including the costs that can potentially be incurred by social demands but that are still not expressed in a product's market price. Underpriced demands can include the cost of emissions reduction to attain high environmental qualities or the option values for natural resources and emissions reduction. LCC includes depreciation of investments, operational costs, allocation of overheads to a product or service (using activity-based costing) and sometimes even infrastructure and related services that are needed to comply with demands. We briefly cover the main types of methods and elaborate on one LCC method in depth. This is the one that aims to translate social demands for emissions reduction into company costs.

Several approaches on LCC can be found, each having several methods. One approach is to value available environmental qualities in monetary terms (contingent valuation). The contingent valuation can be estimated using a few methods. One method is to relate the effects of emissions with the costs of health care, based on valuations of the impact on labor's health (White et al., 1991). Another one relates the effects of emissions on environmental qualities to the willingness to pay for environmental quality (Steen and Ryding, 1993). Still another method links depletion of natural resources with depreciation of the commercial value of natural resources (BSO 1993, 1994). The difficulty in using the contingent valuation is the weak foundation because the valuation of common goods is far from perfect.

The second approach addresses companies' risks connected to regulations and liability that can be imposed by customers and citizens. This includes costs to obtain permits, charges, costs of environmental management, insurance and penalties, as well as the possible costs of a damaged image, disturbed working relations and so on (Savaege and White, 1994; Parker, 1998). The approach received much support in the United States but it is hardly ever used in Europe, possibly because liabilities there presently cover only a very small percentage of all costs connected with environmental quality, whereas an increased risk of liability in the future can only be expected due to a radical change of laws and legal practices, which is uncertain.

Finally, there is a scenario approach that aims to compare current costs with the costs of complying with demanded environmental qualities in physical terms, such as emissions-reduction percentage (Krozer, 1992; Cohan and Gess, 1993; Heijungs, 1994; Duchin and Steenge, 1999; Vögtlander, 2002). The disadvantage is uncertainty about enforcement of the demanded emissions reduction. The major advantage of this methodology in comparison with the others is that the cost figures are real-life assessments based on engineering practices. This makes it possible to validate it for specific companies, branches and regions. The scenarios can be made for all products, albeit the costs of infrastructure can cause disputable depreciation. Hence, the method is close to current managerial practices. The difficulty in calculations is the sensitivity of costs with respect to the demanded emissions reduction because an additional emissions reduction strongly increases costs. However, the problem is solvable, as is shown in the next section, albeit with assumptions. The assumptions are that environmental policies stabilize emissions despite production and consumption growth and that cost-reducing technological progress tends towards the long-term maximum unit costs irrespective of the targeted emission-reduction percentage.

There are many points of criticism about LCC, such as the difficulty in validating the choice of issues addressed in the costing because representation of the demands is disputable. Assessment results are also debated because of poor data transparency and bias caused by outdated figures. There is no simple solution except checking from experience, getting a second opinion and using several other methods. Despite the limitations, LCC is an important tool for life-cycle management by companies because it indicates how to progress. In this way, it supports the development and introduction of innovations that accommodate economical and environmental aims. It also pinpoints at advantages and drawbacks in use of technologies.

7.3 Model for Environmental Strategies

Based on the scenario approach, a life-cycle costing model was developed in co-operation with the Unilever companies to support decision making on the life cycle of products called the Decision Model for Environmental Strategies of Corporations (DESC) (Krozer, 1992). The aim has been to provide cost scenarios in view of demands for far-reaching emissions reduction as described by the model for companies' environmental strategies in chapter 6. The costs to comply with the demanded emissions-reduction targets provide a benchmark for a more innovative, anticipatory life-cycle management. The model is similar to target costing in management accounting, which means that the committed budget, in the DESC model on the compliance strategy is compared with the anticipated efficiency increase at implementation, in the DESC model, the anticipating strategy. The model is elaborated with empirical data.

The demanded emissions reduction is defined for the situation in the Netherlands. The demand is specified for many emissions and expressed as percentage emissions reduction in comparison to the reference situation. The percentage varies from 60% to 95%, which is far beyond current regulations. It is assumed to be an acceptable emissions level in view of health and environmental qualities in Europe (the so-called "no effect level"). The definition of the demands is based on a seminal work by the Netherlands National Institute for Public Health and Environment (RIVM, 1989) that has become a cornerstone in EU environmental policy making. In the report, the demanded emissions reduction is targeted for the year 2010 with the reference year being 1980. The demanded emissions reduction is regularly updated by the RIVM using status reports from the environmental central planning bureau in the Netherlands. The updates indicate that the percentage demanded in the late 1980s is still valid, which means that progress with emissions reduction compensates pollution growth caused by increases in production. This is true, although there are changes in the sector composition such as more emissions reduction is realized in industries compared to agriculture and transport. There are differences between the European countries but generally similarly far-reaching emission reduction is needed to attain the no-effect level and the European environmental policy is progressing towards that goal. Hence, it is assumed in DESC that the reference period for the percentage emissions reduction is the 1980s.

The cost model is founded upon empirical cost functions of emissions reduction technologies that are described in Chapter 3. The unit costs are applicable to many parts of the world because environmental technologies are traded on a global market (Lanjouw and Moody, 1996). It is assumed that the cost functions of emissions-reduction approaches are an asymptote in time due to the technological progress described in Chapter 5. If we consider 3% to 4% a realistic coefficient of the cost-saving technological progress each year and the compliance period of 15 years, the top of almost all cost functions flattens. Note that the assumed technological progress is below the industrial average. The asymptote is an emission-specific maximum unit cost of emissions reduction. The unit costs at the asymptote can be considered the long-term, maximum unit cost irrespective of the demanded emissions-reduction percentage. Hence, on the assumptions of moderate, technological progress and a long compliance period, the model can be used for many situations. The long-term maximum unit costs can be a benchmark for the anticipatory strategy by using cost-reducing innovations (Krozer, 2002). In Appendix C, we present a simplified, exemplary, step-by-step computation of the compliance strategy for biofuel as an alternative to diesel in a private car. This is followed by the data on environmental demands reflecting far-reaching emissions reduction and the long-term maximum unit costs per sector as well as unit costs maximum for all sectors. The maximum unit costs for all sectors aim to indicate proxy-prices of the policies aiming at sustainable development, albeit imperfectly.

For a product, the strategies to comply and anticipate the demanded emissions reduction are assessed in four steps. Firstly, an inventory of the main emissions is made step-by-step in the life cycle, for example, in kg emission per unit of product or service. Then the costs to reduce emissions are calculated based on the technologies that are compiled in the model (compliance strategy). Thirdly, the costs of innovative solutions are simulated by changing an input in the inventory or adding an innovation in the technology database case by case (anticipating strategy). Finally, the costs per step in the life cycle can be summed up to assess the savings made through innovative life-cycle management. We address emissions reduction but assessment of the strategies with respect to health issues, resources of space use, and so on can be done in a similar way. The accounting procedure is shown in Scheme 7.2.

Scheme 7.2. Accounting procedure in DESC model

Activities in the chain		Strict demands		Extra costs of emissions reduction	
Present costs incl. emissions reduction costs	Materials and emissions	Percentage reduction	Emissions reduction in kg	€ /kg reduction	Total costs of emissions reduction
Production ——>					
Distribution ——>					
Use ——>					
Disposal ——>					

The assessments are used to set priorities for environmentally oriented product development (ecodesign) and to test the effects of environmental innovations on life-cycle costs. Priority setting is illustrated by the example of a television's life cycle. Table 7.1 shows the life-cycle costing (Doelman, 1994).

The assessment covers the present costs and the cost to comply with demands. The costs are divided into phases. The production phase includes design, production of components, assembly and packaging. The distribution phase covers overhead of production and sales, delivery to wholesaler, then to retailer and finally to customer. The consumption phase covers electricity use, disposal or packaging and repairs. The disposal phase covers collection and treatment of discharged televisions using landfill and incineration, and recycling as an alternative to the treatment. In the present situation, the largest costs are in distribution because of storage and transport, then in production, consumptive use and finally in disposal. Companies can save money through direct sales to customers or through the Internet to eliminate some distribution costs. The main cost factors change if strict environmental demands are implemented. The highest costs of emissions reduction are in the consumption phase, because electricity costs can increase if power plants have to comply with strict demands, for example related to the climate change policy. Production costs also increase a lot mainly because of the costly heavy metal emissions reduction into water. Additional costs in distribution and disposal are low. The assessment suggests that the cost increase can be reduced by development of energy saving and durable equipment. In production, process-integrated recycling of heavy metals can be cost effective.

Table 7.1. Life-cycle costs of a television in € per set

Steps in the chain [costs of recycling between brackets]	Value in the chain	Additional costs of the demands
Production: design, components, assembling, packing	324	21
Distribution: overhead, transport, wholesale, retail	353	5.5
Consumption: electricity, packaging, repairs	173	49
Disposal: collection, shredding, landfill (alternative is recycling)	6	7.5 [10]
Total	862	83 [86]

7.4 Cases of Life-cycle Management

Below, the results of studies on life-cycle costs are presented. Most studies entailed managerial actions. All studies, except the study on the life-cycle costs of a car, were commissioned by companies (the confidentiality of data does not allow disclosure of all the details). It should be noted that the review illustrate various possibilities for innovations in the life cycle management and possible companies' actions, not to provide a comprehensive of the life cycle impacts of consumers.

The results are summarized in Table 7.2: assessed products (columns 1), units of the products (column 2), consumer price per unit (column 3), extra costs of emissions reduction to comply with far reaching social demands for environmental qualities (column 4), percentage costs of emissions reduction in the unit consumer price (column 5), the main causes of extra emissions reduction costs that indicate priorities for life cycle management (column 6) and finally, the possible decrease of the costs if the main causes are eliminated (column 7). We summarize the goal and scope of the studies, the total costs and main cost factors in the life cycle, potential improvements and actual implementation. The review covers various types of consumers' products. It is a selection of numerous similar assessments purposed to illustrate the variety of social demands for environmental qualities and activities in life cycle management without claiming to represent the overall consumption patterns. The review compiles several studies based on empirical data in the 1990s (Krozer, 1992; Krozer, 1993; Krozer and Lavrano, 1994; Krozer, 1995; Krozer, 1994; Doelman and Krozer, 1995; Doelman and Krozer, 1996; Hennis, et al., 1996).

Table 7.2. Life-cycle costs of products

Products including packaging	Unit	Consumer price, €/unit	Cost to comply with strict demand	Share cost in price	Main cost factor to comply with strict demand	Share main cost factor in total cost
Plant fat	kg	2.3	0.15	4%	Fertiliser use	52%
Tomato	kg	10	0.3	32%	Energy use in production	93%
Animal fat	kg	1.4	0.4	27%	Disposal to sewage	81%
Cotton garment	kg	8.5	0.9	11%	Printing of garments	78%
Washing powder	kg	4.5	0.8	18%	Water treatment	62%
Kitchen block	1 piece	400	30	8%	VOC chipboard	49%
Office chair	1 piece	245	7	3%	Land filling	97%
Copier (excl. paper at use)	1 piece	1 300	114	9%	Electro-technic production	41%
Television	1 piece	925	93	10%	Electricity at use	57%
Car (in km)	130 000	26 288	3 708	14%	Fuel use	93%

Plant fat. The assessment was done to define priorities in life-cycle management. The life cycle covers agricultural production, industrial processing, packaging, transport, use and waste processing. The share of compliance costs in life-cycle costs is about 4%. Fertiliser emissions on fields are the main cost factor (52% of compliance costs). The life-cycle cost is typical for many high-value food products. The main cost factor is in agriculture, but the total compliance costs are low. The effects of improvements are also assessed. The focus is on sunflower farming. The main proposed improvement is balanced fertilization and integrated pest management to reduce emissions to soil and water. It can be implemented, but at a high production cost in agriculture. A few years after the study was done, the company started implementing pilots to upgrade the supplied agricultural resources in various countries.

Greenhouse tomatoes. The goal was to support the users of processed tomatoes for ready-to-eat meals. The assessment covered cultivation and wholesale of tomatoes in greenhouses compared to open-air cultivation. Compliance costs covered about 32% of life-cycle costs. The main cost factor in greenhouses is CO_2

emissions due to intensive energy use. Energy saving can reduce the costs by 93%. The second cost factor is substrate that becomes hazardous waste after use. Open-air cultivation is cheaper but it is seasonal. This life cycle is typical for greenhouse products. The main improvement is energy management, e.g. utilization of waste heat, albeit the benefit is situation specific. It is also possible to focus production on high-value products. Integrated pest control reduces waste. The assessment contributed to the supply chain management. Some focial environmental issues were implemented in quality standards.

Animal fat. The goal of the study was to assist management in setting priorities. The assessment covers the life cycle of fat: fishing and fish processing, oil crops, industrial processing to fat, packaging, transport, consumption and disposal. Compliance costs cover about 27% of life-cycle costs. The major cost factor is disposal of fat to sewage after usage (about 80% of the compliance costs) because fat clogs pipes and it needs heavy cleansing at treatment plants. This life cycle is typical for animal fats. A major improvement can be made if consumers dispose old fat in bins for organic waste instead of in the sewage. The change does not require an investment, but depends on consumer behavior. The company checked the outcomes in co-operation with a consumer panel and found a highly positive response. Thereafter, using promotion, the company fostered consumer awareness to drop fats into bins for organic waste.

Man's shirt. The study aimed to foster environment-oriented procurement of a retailer. The assessment covered the main steps in the life cycle: cotton cultivation; weaving; printing and production of a cotton shirt. Compliance costs cover about 11% of the life-cycle costs of a shirt. The main cost factor is printing the shirt because of the heavy pollution released into water (78% of compliance costs). This is typical for many products treated with toxic chemicals. The main improvements are the use of low-toxic dyes and modern printing technology. Although environmentally friendly alternatives are more expensive than regular ones, the combination of sound machinery, prudent resource use and higher-value sales provides opportunities for the low cost or even beneficial change-over. The assessment confirmed the retailer's expectation about opportunities. Strict procurement specifications for suppliers have been implemented.

Washing powder. A study was done to support product development and engineering of washing powder. The assessment covered the main steps in the production of ingredients for washing powder: manufacturing, packaging, use and disposal to sewage. Compliance costs cover about 18% of life-cycle costs. The main cost factor is connected with treatment of wastewater after use, mainly because of a few poorly degradable compounds in washing powder (62% of compliance costs). This is typical for many hygiene and personal-care products. The main improvement is a more degradable product that is more costly; and better dosage of powders by consumers that saves money in use. After some additional studies, the producer adapted the composition of washing powders and packaging with a net benefit.

Changes in washing powder composition are simulated to illustrate the effects of product innovations on the costs of emissions reduction. The simulations are done with fictitious data, but are close to reality. Four alternatives are assessed: phosphate powders, washing powders with sulfates and zeolites as substitutes for phosphates, compact packaging and biodegradable washing powders. The speed of innovations was impressive; all these changes were still being tested in laboratories at the start of the study in 1991. All of them had been introduced five years later. Graph 5.4 shows the costs of emissions reduction to attain strict demands in the life cycle of products. The costs of washing powders with phosphate could increase as much as € 1.90 per kg because of extra costs in the production of phosphate (mainly because of costly heavy-metals emissions). The substitution of phosphates with zeolites largely reduced the potential costs. The costs in use became the main high-cost factor, because of costly degradation at water-treatment plants. Changes of packaging and better degradability reduced the potential costs of emissions reduction to less that € 0.50 per kg.

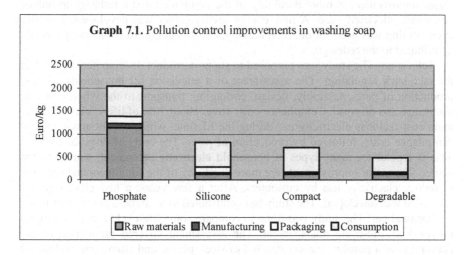

Graph 7.1. Pollution control improvements in washing soap

Kitchen block. A study was conducted to redesign a kitchen block. The assessment covered production of all the main parts needed for the kitchen block: chipboard, glue, metals and plastics, assembly of the block, use and disposal. The chipboard is the largest part. Compliance costs are about 8% of the life-cycle costs. VOC emissions during manufacturing of the chipboard are the main cost factor (about 49% of the total compliance costs), followed by disposal after use (adding up to 94% of compliance costs). This is typical for furniture. Improvements include the use of higher-grade, more durable materials, low glue use and recyclable constructions. The cost of the newly designed kitchen block has increased but with higher performance at use and durability. The company has manufactured a series of new kitchen blocks, but sales for the lowest-price segment were low. Hence, a redesign was made for the higher-price segments with reasonable sales.

Office armchair. The study was carried out in support of product development, in particular, addressing the take-back and recycling of products after disposal. The assessment covers production of the main components from plastic, steel and aluminum production, manufacturing with coating of the armchair, use and disposal. Compliance costs cover about 3% of the life-cycle costs. The main cost factors are VOC emissions during production of plastics, not at manufacturing because the manufacturer has installed state-of-the-art technology for low-VOC powder coating and the landfill costs after use. These two cover about 97% of compliance costs (similar to the results of the kitchen block). Further studies on the feasibility of take-back and recycling schemes have shown that take-back and recycling are cheaper than disposal to landfill. A redesign of armchairs with take-back possibility followed and has sold successfully.

Copier. The study was made to support product development of a professional copier. The assessment covered production of the copier: manufacturing of the main parts, assembly and use by customers, including maintenance (but excluding paper). Compliance costs cover about 9% of life-cycle costs. The main cost factor is production of electronic parts that comprises about 41% of compliance costs, followed by electricity during usage. This is typical for electronic equipment. The major improvement is more durability of the equipment and a redesign to reduce customers' electricity use. A new energy-saving copier was developed. The life-cycle costing was used for decision making, but it is unknown if the study directly contributed to the redesign.

Television. The study was proposed to support product development in view of the take-back regulation. The assessment of a television for households covered production of parts, assembly, design, packaging, transport to the wholesaler and shops, use and disposal. Compliance costs cover about 10% of life-cycle costs. The main cost factor is electricity use during the lifetime, which covers about 57% of compliance costs, followed by production of parts. The costs of disposal are low. This is typical for many types of household electronic equipment, similar to the copier. The major improvements are more durability of the equipment and redesign to reduce electricity use by customers. After a few years, a low electricity use television was developed. The study has contributed to awareness about priorities.

Personal car. The study was done to compare some life-cycle methods without intention of life-cycle management. The assessment covered production of the main parts of a gasoline car constructed of steel, plastic and aluminum, and use of a car for 130 000 km of driving. Compliance costs are about 14% of life-cycle costs. The main cost factor is the consumption of gasoline that covers about 93% of compliance costs. This is typical for transportation means. The main improvement is fuel saving during use. The study is used for life-cycle management of cars. After the study, it was assessed how to introduce a policy on car inspections that could reduce gasoline consumption. The assessment compared the effects of annual maintenance and inspection schemes with the system of quality assurance by consumer organizations and dealers on the consumption of gasoline. Several arguments favored the quality-assurance scheme, such as lower costs and lower gasoline use. The results were not used because authorities preferred annual inspections.

The presented case studies on life-cycle management show that stricter environmental demands invoke higher costs. The use of available technologies to comply with strict environmental demands would increase life-cycle costs by 3% to 32% in comparison with moderate demands. Consumer prices at the end of the chain would necessarily increase much more because every step in the chain must add a percentage, an additional profit margin to the cost in order to stay in business. The use of innovations in the life cycle enables life-cycle management to mitigate the cost increase.

7.5 Conclusion

The question of how companies respond to demands for environmentally sound products and services is discussed with reference to life-cycle management. A major difficulty in companies' decision making is that social demands such as the demand for better environmental performance of using products during their life cycle, often contradict private demands that are demands for better product functionality. Innovating companies, therefore, must deliberate trade-offs between product development to satisfy private and social demands. Innovators must also consider that social demands are usually not incorporated in market prices nowadays but they can be incorporated in the future using policy interventions such as charges and subsidies, as well as a growing consumer readiness to spend money on environmentally sound products. To cope with the trade-off, many methods have been developed to assess the economic and environmental impact. Life-cycle costing is becoming widely used among the methods that link environmental and economic performance in the life cycle. Life-cycle costing is, in essence, an assessment of present and possibly future costs connected with private and social demands. Based on life-cycle costing, scenarios can be drafted to indicate the costs and benefits of improvements in the life cycle.

The cases of improvements in the life cycle illustrate how strikingly innovative life-cycle management can be. The results indicate that companies often benefit from anticipation of social demands, particularly through lower life-cycle costs in use. It is also found that the costs of far-reaching emissions reduction strongly differ between products. The costs can increase anywhere from 3% up to as much as 35% of the consumer price if only available technologies are used. Thus, demand for far-reaching emissions reduction can increase market prices in comparison to moderate demands under low-innovative conditions. If the response is low-innovative, then enforcement of strict demands causes negative productivity effects. The presented cases, however, illustrate that a few focused innovative actions can counteract this and bring the price increase down to a low level. Some actions are even net beneficial to the producer because they can reduce inefficiencies in the life cycle, albeit an initial investment is needed.

Life-cycle management has an advantage above the engineering focus on manufacturing alone. The cost-saving actions can be identified in every step of the chain, thus by dedicated trade to provide resources and materials for products, by additional services to customers, product development, in marketing and so on. All of these are in addition to innovations in the manufacturing phase. Some actions are risky because of the large costs of development without a guarantee of success. However, many other actions are low-risk improvements in promotion and supplies. The disadvantage of life-cycle management is complexity in decision making because actions depend on different interests in the chain. Selection of the key cost factors and innovations determines the costs and benefits. A few focused innovations that substitute the available technologies reduce the life-cycle costs substantially.

The major difficulty in the life cycle management is that innovative, beneficial actions can only be assessed case by case, because the life cycles of products diverge too much to allow generalizations. A distinction in life-cycle management should be made between short-cycle and durable products. The highest costs of compliance with the social demands in cases of short-cycle products are often found in the supplies (e.g. agriculture in food) and in disposal (e.g. packaging). Conversely, the highest compliance costs in cases of durable products are usually found in the manufacturing of components and during product use, e.g. electricity and fuel use. For this reason, producers of short-cycle products benefit mainly from the better selection of suppliers. This means that suppliers must be able to accommodate strict social demands at low cost. It is quite different for producers of durable products who benefit mainly from the development of more durable and energy-saving products. This development can cause a high purchase price, but lower costs at use. Producers of durable goods face the challenge of convincing customers about the benefit of qualitative products in the life cycle.

8

Environmental Policy for Innovations[8]

The question is: Can policy makers prepare and enforce demands in such a way that companies are able to develop new technologies? And if so, how? Environmental policy is discussed from the perspective of technology developers, the innovators. It is shown that uncertainty about policy preparation and enforcement are the key factors involved in the supply of environmental innovations.

8.1 Induced Innovations

The prevailing view about development of environmental technologies found in handbooks on environmental economics and management is that policy makers prepare regulations to enforce demands that aim to maintain good environmental qualities.[9] In this way, the policies create a market for environmental technologies because enforcement of strict demands pushes companies to search for compliance at the lowest costs. If there is no suitable technology, then polluters develop a new technology or they pay other firms for its development. Within the theory, there is debate about what type of instrument is the most effective one to invoke innovations. Is it direct regulations through permits, market-based regulation using policy charges, subsidies and emission trade, or is it social regulation undertaken through agreements? Thus, it is assumed that policy makers invoke environmental innovations. However, the validity of instrument theory holds only under the assumption that policy makers can prepare and enforce such strict demands that new technologies must be developed and that the development can flexibly be

---

[8] Parts of this chapter were published in Krozer Y., A. Nentjes, (2006), Environmental Policy and Innovations, *Business Strategies and the Environment*, published on line DOI 10.1002/bse. I am grateful to Wiley Interscience for permission to use the material.

[9] Look for examples in Pearce and Turner, 1990, p. 61–119, Tietenberg, 1994, p. 211–277, Perman et al., 1996, p. 216–250, Reijnders, 1996, 15–24.

adapted to all demands. The assumptions are neither realistic nor theoretically valuable starting points.

The view that policy makers can implement the standards that invoke technology development is not realistic because the ALARA principle (as low as reasonably achievable) in environmental policy presumes the availability of BAT that are not excessively costly at emission sources. The demands that go beyond BAT are realistic only for the so-called "technology-forcing standards". Such technology-forcing standards have hardly ever been set because of the risk that policy implementation fails, which damages the authority's reputation in cases where non-compliance must be tolerated because suitable technologies were not developed in time, or have not functioned properly. One such case is the "Zero Automotive Emission" regulation in California starting in 1990. It prescribed such low exhaust emissions for cars that only experimental electric or hybrid fuel-electric cars could attain them. Its implementation has been postponed several times because of poor technologies (Calef and Goble, 2005). Another example is the "Verpackungsverordnung" in Germany beginning in 1988, which has required 80% to 90% recycling of packaging waste without consideration of available recycling technologies, but many exemptions had to be allowed because the targets could not be attained. The theoretical problem with the assumption is that technology-forcing standards are not enforceable. So, if a suitable technology for environmental demands is not available then policy must tolerate disobedience, which in turn undermines the legitimacy of the authority. Hence, technology forcing is risky. The perceived risk is so high that it is seldom used. The risk is real; note that precisely because of the lack of suitable technologies, technology-forcing standards were abandoned in both examples. Hence, policy makers can force technology only in exceptionally important cases with the assurance of technology developers that a suitable technology can be provided some years after the announcement of technology-forcing standards.

The second assumption in the instrument theory is that environmental technologies can flexibly be adapted to stricter demands. This assumption must be dismissed in view of the cost-engineering functions of emissions reduction discussed in Chapter 3. It is shown that process variables of emission sources determine the costs and effects of environmental technologies, so the supply of environmental technologies for various sources can be deficient. Hence, it is realistic to assume that innovators enable polluting companies to comply with some environmental demands at specific sources. The technology is developed if the innovators expect sufficient revenues from sales to make a profit and the users of innovation expect to generate benefit during use. This is empirically underpinned in Chapter 4.

Another point of view is that innovations enable preparation and enforcement of stricter environmental demands. Generally, environmental demands are determined by available environmental technologies and development of new ones. Thus, the rate of technological progress determines how strict the demands can be that will be implemented, not the other way around. The issue is: What can policy makers do to enhance progress? This is addressed by linking instrument theory with innovation theory that is introduced in Chapter 4. Innovation theory is focused on the innovation rent that is the sum of expected profits of innovators and benefits of users. Following innovation theory, the profitability of innovations is determined by four factors: the investment costs to realize a new technology, the expected sales revenue, the uncertainty about sales of the innovation and the lead time to market between the investment and sales. We argue that the profitability of technology development is mainly influenced by the lead time and uncertainty in environmental policy.

Firstly, we discuss instrument theory on technology development and review instruments in environmental policy from the innovation perspective. Then, we link instrument theory and innovation theory in a model for environmental innovations. Instrument theory provides insight into the effects of policy instruments on innovations, whereas innovation theory explains the factors that influence innovators' decisions. The model is used to assess the profitability of technology development.

8.2 Instrument Theories and Technological Development

There is a huge body of literature about the relationship between instruments in environmental policy and innovations. It is briefly reviewed here. In the theory, the perspective of a policy maker is taken that is seen as an authority who prepares and enforces strict policy demands to reduce emissions. The effects of policy instruments on environmental innovations are analyzed in view of the demanded emissions-reduction percentage. It is assumed that policy makers can choose any instrument that is sufficiently strong to attain the demanded emissions reduction. The basic choice is between direct regulation (command and control) that is based on emissions standards in permits for production license and market-based regulations using economic instruments such as the emissions charge (Downing and White, 1986; Nentjes and Wiersma, 1987). The assumption is that companies implement additional technologies as long as they reduce costs (at the margin). It is argued that market-based regulation gives a stronger incentive for innovations than direct regulations. The argument is that economic instruments stimulate cost-reducing and effect-increasing technologies because residual emissions have a price, whereas emission standards lack incentives to reduce emissions beyond the standard.

The argumentation is presented in Figure 8.1. The unit costs of emissions reduction are shown vertically, the emissions-reduction percentage is given horizontally. If a company must comply with a policy demand for emissions reduction (depicted by a shift from Er_1 to Er_2) then it can use available technologies at the unit costs Cr_1 or apply innovations that reduce the unit costs to Cr_2. For a given demand, the field ABCD shows savings due to innovations under direct regulation, whereas the field ABED shows the savings gained under market-based regulation by an emissions price P. The extra incentive to innovate under market-based regulation is the equivalent of the field CDE.

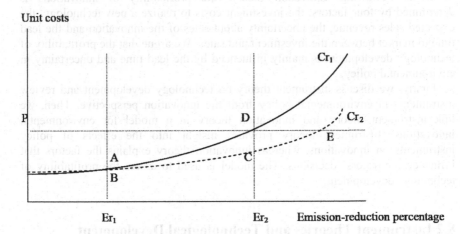

Figure 8.1. Theory on the effects of direct and market-based regulations on environmental innovations

This argumentation is fine tuned. Below, we address the main nuances in instrument theory that incorporate theories on technology development and institutional theory. Thereafter, the effects of the main policy instruments on environmental innovations are discussed based on policy practice.

Several studies on technology development distinguish between decision making in polluting companies and in specialized departments or in specialized companies for environmental technology. Similar to the joint production function for production and environmental demands, the studies assume allocation of R&D expenditures for the output or for the environment. The former contributes to profit, the latter to compliance with environmental policy. It is assumed that R&D expenditures are equally effective and that economic instruments like charges provide more flexibility in the allocation of R&D efforts than emission standards because the latter obliges companies to spend on emissions reduction. The analyses suggest that the effect of policy instruments on the development of environmental technology depends on the substitution–elasticity between R&D options. If the substitution is elastic (elasticity above 1) then the charges more strongly encourage R&D for environmental technologies than the emission standard. However, in cases of inelastic substitution, permits stimulate R&D for environmental technology more than charges (Magat, 1978). The argument used in support of inelastic substitution, hence in favor of direct regulation by permits as a more effective instrument for environmental innovations, is that environmental policies are uncertain so it is unattractive to invest R&D in environmental technology (Mendelsohn, 1984). Inelastic substitution is also expected in cases of costly technology development and in cases of small-scale emissions like small companies (Becker et al., 1993). It is also argued that the combination of permits with economic instruments is most effective, whereas the comparison of a single instrument like subsidies, charges and tradeable permits is inconclusive because it depends on policy specifications (Milliman and Prince, 1989). The arguments derived from evolutionary distinguish between the add-on and process-integrated ("cleaner") technology. The arguments support the view that combinations of instruments are needed. It is argued that market-based instruments provide more flexibility to companies, so it is easier to develop and implement cleaner technologies (Verbruggen, 1991; Kemp, 1995).

Studies from the institutional perspective suggest that the policy framework largely determines the effects of policy instruments on technology development. If the revenues of charges are not given back to the companies, then companies pay for the charges and the technology at a higher cost than in the case of a permit. Hence, the charges constrain budgets for development of environmental technologies. The policies can also have different goals, so the instruments can cause contradictory effects, e.g. the goal to generate income instead of emissions reduction can provide charges that do not stimulate development of environmental technology (Bohm and Russel, 1985). The point is also made that the theory often assumes stricter demands for emissions reduction, but less strict demands strongly decrease the positive effects of economic instruments on technology development. This is because R&D expenditures on environmental technology become uncertain. The incentive to reduce cost remains under the permit (Heyes and Liston-Heyes, 1997). It is also shown that the specifications of economic instruments have different effects on technologies (Opschoor and Vos, 1989; Kip and Krozer, 1990, Barde, 1999). The effects depend on the basis for the instrument, fee rate, scope of sources, input or output orientation. A variety of instruments is noted. There are emission charges, user charges for services, administrative charges for registration, product charges to invoke substitution of products, variable taxation and pricing, VAT regimes, tariff systems and more. Many types of subsidies are found, like grants, soft loans, guarantees, participations as well as indirect support by tax exemption and allowances. The subsidies can be for technology development or for emissions reduction. One finds deposit refunds with variants for products, materials, with and without VAT subtraction. Finally, there are market-creating instruments such as emission trading, user rights, quotas, price and quantity guarantees, certificates with liabilities and emission bonds.

Although the comments do not abolish the foundations of instrument theory on technology development, there are so many nuances that the policy instruments must be specified with respect to the conditions that favor innovations. We address emission standards in permits, charges, subsidies, tradable rights and covenants. The emission standard in permits can be tuned to emission sources. Such a differentiated emission standard is used as a reference to compare the effects of instruments on development of environmental technologies.

Emission standards. In the theory, emission standards in permits are usually analyzed, whereas technology standards in permits are widely used in practice. Technology standards specify the technological means, like the maximum pressure in boilers for safety reasons or concrete floors at gasoline stations to protect soil. The performance of technology is usually prescribed in detail, sometimes the type of manufactured equipment or even the manufacturer. Emission standards specify the concentration of emissions at the source, but in practice, the handbooks for enforcement specify the BAT for every type of emission source. Hence, the suppliers of the BAT can get a dominant market position and hold it for many years until a new BAT list is approved. This impedes innovations.

Contrary to the technology standards that are usually uniform at many different sources, emission standards are often adjusted for various sources. Although the standards are usually based on international or national directives, translation into regional policies and enforcement of the standards is usually adapted to the specific regional and sector situations. For example, it can be a waiver for non-compliance. This means that policy makers tolerate incompliance for a number of years to enable technology development or to make an adjustment in the economic structure. It is increasingly common for authorities and companies to agree on the broad lines for emissions reduction in the permit and leave the implementation to the companies, which enables them to innovate. Thus, enforcement is increasingly a negotiation process about the minimum standard in the short-term and advancing emissions reduction in the longer term (Doelman et al., 1991; Kemp, 1998; Kloppenburg, 2000). As a result, emission standards differ. The differentiation reduces effectiveness, because less strict demands are enforced at some sources that potentially can reduce more emissions, but adjusting the standards to companies' situations increases efficiency. Case studies on the costs of reducing heavy metals released into water (Klink et al. 1991) and biodegradable matter to water (McConnel and Schwartz, 1993) indicate that the differentiated standards are comparably efficient to the charges. Static efficiency is similar.

Charges. Studies show that high charges on emission sources are effective (Bressers, 1983; Schuurman, 1988). It is often assumed in instrument theory that the charges regulate emissions (regulatory charges), but the charges are usually meant to finance public policy, for example, construction of public water treatment (financing charges). The effects of both can be similar if companies can reduce emissions at the source, which is an unintended effect of the financing charges, but usually the financing charges are put in such a way that the regulatory effect is avoided. The regulatory action cause unintended effects. It can be expected that the effects of the financing charges on environmental innovations are low and even negative because the charges divert expenditures from emissions reduction at the sources to public investments or finances. A regulatory charge has two positive effects in comparison with the differentiated emissions standard. Firstly, the total market volume for environmental technology is larger because the residual emission is charged. Secondly, the use of environmental technology depends only on costs at the source, not on other policy objectives, which reduces uncertainty about R&D expenditures. However, regulatory charges also have negative side effects on innovations. It is found that the high administrative charges for tests and precautionary measures for medicines and pesticides in Sweden unintentionally causes a barrier for entry of environmentally sound substitutes of products and technologies (Fleischner, 1998). High administrative charges on landfills in densely populated areas of the United States limit landfill in these areas, but cause more waste to be transported to less-populated areas with less monitored landfill. The basis of the charge largely determines its effect on environmental technologies. For example, if the user charge on waste is set per household then it hardly influences waste prevention because there is no incentive to reduce waste. However, a user charge per volume of waste triggers waste prevention because it provides savings, albeit negative side effects of waste transport need to be prevented (Kip and Krozer 1990).

Subsidies. It is a common practice that the development of environmental technologies is heavily subsidized, usually to strengthen local industries. The form of subsidies differs by country and sector, but usually more than three-quarter of development costs are funded with public money. There are three types of subsidies for environmental innovations: subsidies for developers (innovators), subsidies for procurement and subsidies for emissions reduction. Subsidies' effects on environmental innovation is discussed in comparison with the differentiated emission standard, but the discussion is necessarily rather basic because each type of subsidy entails many variants. For example, subsidies for innovators can address know-how and information transfer (missions, fairs, conferences), funds for different phases of technology development (research, development and demonstration), sector-specific support (e.g. energy saving) and risk reduction of financiers (e.g. exemption of green investments from dividend tax).

The subsidy for the technology developer (innovator) is given to realize a new technology. The objective can be to strengthen technology suppliers and to enable industries to comply with future, stricter, emission standards at low costs. The latter type is not meant for technology development as such, but rather to support policy making that aims to enforce a stricter emission standard. We discuss only the latter. The subsidies' effects on environmental innovations are analyzed based on a model of a developer who can decide to develop a more effective technology in view of policy making (Nentjes, 1988). In the model, stricter standards enlarge the market, but the chance of enforcing stricter standards decreases. The model shows that the subsidy has two positive effects: It reduces the costs of technology development and it increases the chance that the emission standard is tuned to the subsidized new technology. In the model, the subsidy is strictly targeted at the development of a more effective technology, but in practice the subsidies are often provided for various goals such as export to countries with standards that are lower than the domestic standards, which can distort technology development in the host and recipient countries. Moreover, those evaluating a proposal for subsidies are usually policy makers and experts with vested interests who are inclined to support their interests.

Subsidies for the procurement of new environmental technologies are mainly aimed at reducing investment costs at the source. The argumentation in favor of this type of subsidy is that lower investment costs stimulate procurement of cleaner technologies. Hence, sales of cleaner technologies increase, that in turn encourage technology development. Empirical findings based on the allowance of fast depreciation of cleaner investments in the Netherlands (VAMIL) revealed that administrative imperfections distort the positive effects of procurement subsidies on innovations. For example, the list of cleaner technologies lags way behind the availability of new technologies (Nentjes and Scholten, 1989). The clarification of the list with new technologies is administratively complex, because various interests press for the status quo. Hence, the list has hardly changed in ten years, even after legal renewal of this regulation. The consequence is that procurement of available technologies is supported, whereas the sales of new ones are reduced. This hinders innovations. Innovation is impeded if the available technologies become cheaper due to the subsidies and the sales price of available technologies decreases in comparison with the price of new technologies.

Subsidies for emissions reduction can be per unit emissions or per unit emissions-reduction costs. The first type is introduced within the framework of the Kyoto agreement on the reduction of greenhouse gas emissions by the so-called Joint Implementation and Clean Development Mechanism. This means that the countries that ratified the agreement can get financial support for equivalent CO_2 reduction. The second type is, for example, subsidy for sales of electricity based on renewable-energy production. This type of subsidies was introduced in several European countries and gained support of the producers. The subsidies have fostered diffusion of these technologies due to enlargement of the market for renewable-energy technologies but effects of such subsidies on technology development are less clear. It is also advocated to entrench this type of subsidy to support the costs of emissions reduction. This is done on the argument that the subsidy (as opposite of a charge) can prevent the negative effects of emissions reduction on international competition. The argument is that subsidies have a similar effect on emissions reduction as charges if information about costs and effects is available. A critique is that imperfect information causes unfair competition because some sources can receive too little and others too much subsidy in comparison with the untreated emissions. This argument is counteracted by advocating a premium for public information that should increase at lower emissions per output, because it becomes more costly to reduce emissions (Carraro and Sinicalco, 1992). However, it is unclear how to verify the reliability of the information that is provided to the public. There are also several other arguments against subsidization per emission or per unit cost. The administration is troublesome and causes inefficiencies, because emissions reduction is difficult to monitor let alone the unit costs of emissions reduction. Subsidies also cause price distortion because they benefit the growth of polluting production by reducing the production costs, thereby causing unnecessary extra growth of emissions (Kanazawa, 1994; Pieters, 1997). In addition, it can be argued that the largest sources receive the most subsidies, thus distorting the scale effects. Therefore, it is advocated to combine charges with subsidies for procurement of technology, which is a variant of the deposit-refund system (Carraro and Siniscalco, 1994; Sigman, 1995; Palmer and Sigman, 1997). It can be effective, but it is costly to administer, because charges and subsidies must be registered. This instrument has been applied for dissemination of a three-way catalyst in cars in the Netherlands with moderate success, but there is no experience with effects of such subsidies on technology development.

It is difficult to compare subsidies' effect on environmental innovation with the effect of a differentiated emissions standard because much depends on the specification of the subsidies and the way they are administered. If the administration is effective and efficient then the subsidies tend to be effective as well, in particular, subsidies for technology development. However, in general, administration is far from perfect so subsidies can create a barrier for innovations.

Tradeable rights. The instrument of tradeable rights is based on the assumption that the rights for environmental qualities can be divided among private interests and that negotiation can balance the interests of polluters and those who are harmed. The instrument is first presented as a right of inhabitants to buy and possess environmental qualities (amenity rights). The proposal is made to strengthen the rights of harmed groups confronting the polluters (Mishan, 1993). The instrument is used in nature management, for example, management of some national parks by environmental organizations, but it is not really developed as a general policy instrument. The system of tradeable emission rights has broader applications. The system of tradeable emission rights means that policy makers set an emission maximum in a region or sector (emission ceiling), divide the allowed emissions volume between polluters as a right and allow trade between the emission sources (transactions with the rights). Positive effects of tradeable emission rights on technological development are expected, because polluters must buy equivalents of emissions reduction to produce for growth or to compensate emissions growth by progress in environmental technologies (Dales, 1972; Tietenberg, 1984, 1991). Tradeable emission rights were introduced in 1995 in the United States, using SO_2 emissions in the electrical power sector. In this case, the emission ceiling was set for a number of years and the rights were divided between the power plants in such a way that the rights equaled the total emissions allowed under the ceiling. The sector and external organizations could buy and sell the rights. Companies that did not use the rights to cover their emissions could store the rights at the bank. The positive effects of tradeable emissions rights on development of environmental technologies are disputed. This is done using the argument that the technologies are not divisible, that companies make high transaction costs at the expense of technology development and that the companies can trade rights in order to inhibit the entry of more innovative companies (Malueg, 1989; Marin, 1991). However, empirical findings do not confirm these objections. The costs of SO_2 emissions reduction in the electrical power sector in 1995 were reduced in comparison with the costs in 1990, mainly due to the substitution of high-sulfur coal by low-sulphur coal (about 40% of the total cost decrease), by less transport of coal (about 50% of the total cost decrease) and by better mixing of coal. The costs of treatment were also reduced by better capacity utilization. Thus, there are efficiency advantages in comparison with emissions standards, albeit little technological development is found so far (Ellerman et al., 1997:44; Klaassen and Nentjes, 1997:399).

Covenants. Covenants are voluntary, private agreements between authorities and industries on emissions reduction. Covenants started in the mid-1980s in the Netherlands. They became a cornerstone of Dutch policy with 80 covenants in place by the early 1990s (van de Meer, 1997). Recently, covenants became a major policy instrument in the EU. The number has grown from 23 agreements in the period 1982–1986 to 123 in the period 1992–1996 (Karamanos, 2001). The main advantage covenants offer companies is that they can negotiate less-strict agreements, and thus postpone stricter, direct regulation (Wagner, 1991). This advantage is also suggested by the study on the international covenant on 50% CFC reduction (Montreal Protocol). The goal for the CFC reduction was attained in a few years time at low costs because companies in the countries that ratified the covenant could reduce emissions even before signing the agreement. This was due to available substitutes for many CFCs (Murdoch and Sandler, 1997). The main reason for authorities to introduce covenants is to shorten the lead time in policy making as preparation of a covenant is less laborious than regulations. In reality, evaluation of eight Dutch covenants signed in the 1980s showed that the preparation time of covenants strongly varies from 0.5 to 7.5 years. The average is about six years. Compliance is high in four out of eight cases (Klok, 1989), which is confirmed by the study on six covenants in the EU (European Environmental Agency, 1997). A study that examined regulations and covenants on the production of VOC-free paints by the paint industry and prevention of cadmium in products in the Netherlands suggests that companies are not in full compliance with either one. Covenants have greater non-compliance than regulation because only 60% of companies have implemented the agreement and only a small minority has done so fully. It is also found that the technological and economic feasibility of environmental technologies determine compliance. Compliance improves if low-cost technologies are available (Van Peppel, 1995). The effectiveness of covenants in comparison with emission standards is also disputed because a large part of compliance is not done voluntarily, but rather is enforced through permits (Wit et al., 1999). Covenants' effects on technological development are controversial. Some authors expect covenants to provide a main policy instrument on technological development because the choice is left to companies (Wallace, 1995). An opposite view is that covenants delay technological development because of piecemeal compliance (Sunnevåg, 1998). Experiences with waivers for compliance with environmental regulations in the United States in the 1970s, confirm the latter. The waivers did not stimulate innovations, because it was possible to delay compliance with agreements (Ashford and Heaton 1979; Ashford et al., 1985). Similarly is concluded in the study on various negotiation oriented policy instruments instead of regulation in the United States, such as provision of information to the public and authorities, provision of technology reviews by authorities and to companies and agreement about companies' plans (Ashford, 1996). It is also found in a study on six cases of permits based on companies' plans for emissions reduction instead of emissions standards. The agreements about flexibility in the permit did not stimulate innovations and in some cases they provided opportunities to restart negotiations about regulations (van de Woerd, 1997). Opinions confirm the finding. The opinions of 14 companies and of six external experts on the effects of covenants on technology development in three branch-wide covenants (printing industry, dairies and metal product industry) are strikingly different. The experts value the effects of covenants on technology as much higher (average 7) than the companies (average 2). It is also found that enforced technology agreements help. For example, in the covenant with the printing industry, it is agreed to define the new technologies in manuals every four years. The result was that selected, new technologies are immediately recorded in 30% of companies' manuals and in 50% of companies with some delay (Hofman and Schrama, 1999). The studies suggest that covenants have hardly any positive effects on innovations; the effects are less positive than the differentiated emissions standards. The emissions reduction in covenants is lower than in standards and the rate of compliance with the agreed emissions reduction is slower too.

Based on instrument theory, it can be argued that economic instruments provide stronger incentives to innovate than differentiated emissions standards, but also that the positive effects depend on the specific design and enforcement of the economic instruments. The advantages of the economic instruments for development and diffusion of environmental technologies in comparison with a differentiated emissions standard are conditional. On the assumption that polluters have no spontaneous interest in emissions reduction, the following main conditions for innovation can be summarized. The positive effects of charges on innovations are in cases where the collected revenues are allocated for development and implementation of environmental technologies and there are low administration costs. Subsidies for technology development and procurement of technologies alone are less effective than emissions standards. Subsidies in combination with emissions standards can trigger innovations if administration is effective. However, if the subsidies are largely provided to available technologies then they impede environmental innovations because the price of the available ones is reduced. The subsidy per unit of emissions reduction is not a mirror of charges because of two negative effects the subsidy has on the costs of emissions reduction: the administration costs are high because information about cost functions of emissions reduction is needed to allocate subsidies fairly and the effectiveness is low because the growth of emissions sources is subsidized. Tradeable emissions have positive effects on environmental innovations if the ceiling for emissions is low. The effect of enforcing tradeable emissions on innovations is similar to emission charges, but without the negative side effect of diverting private funds for public goals. The implementation of environmental technologies can be fast because companies have an interest in reducing emissions in order to grow. Covenants have a positive effect on innovations in the case of the threat of direct regulations for non-compliance and in cases of subsidies in support of the covenant. Otherwise, covenants provide no incentives for innovations unless there is liability for non-compliance.

The limitations of instrument theory are the assumptions that new technologies can be forced by regulations and that companies react automatically to policy instruments, which happens only sporadically. Instrument theory partially explains the dissemination of available technologies from the past with possible adaptations because specific policy instruments influence the market volume for the technologies. In this respect, charges and tradable emission rights can be advantageous. However, instrument theory does not explain innovations and the first stages of diffusion, because it does not consider R&D's high costs of technology development that must be made to generate sales revenues several years later. The key factor is innovators' expectations about the possibilities to develop a new technology and to cover the costs through sales revenue.

To develop and sell environmental technologies, innovators must consider the process of policy making. The process that is often called a policy cycle, starts with the signaling of an environmental problem. It entails preparation and enforcement of regulations and ends with evaluation of implementation of environmental technologies by companies at emission sources. The policy cycle can take a few decades. The difficulty for any innovator is that the policy cycle's duration is extremely long and the results are uncertain. There are phases in the policy cycle that can be related to the innovation cycle that embraces development and dissemination of environmental technologies.

8.3 Policy and Innovation Cycles

8.3.1 Policy Cycle
a. Signaling period
A policy cycle usually starts with the signal of degradation of environmental qualities by pollution such as the negative effects of emissions on safety and health. The effects are signaled by interests such as environmental groups, local communities, experts or companies. Only some signals are sufficiently strong to trigger policy preparation. It also takes time to build social and political pressure that is necessary to trigger policy preparation. The period between the signaling and the start of policy preparation usually takes more than a decade. For example, the negative effects of pesticides on the environment and health were signaled by scientists, farmers and workers in 1946. For more than the next 30 years, research was done to underpin the necessity of policy in this field, whereas policy preparation only started in the early 1970s. This signaling period was more than 30 years (Sheail, 1991). The signaling period of ozone depletion was about 15 years, which is rather short in environmental policies. Ozone depletion by CFCs was signaled by scientists in 1972, whereas policy preparation for CFC emissions reduction started in the mid-1980s. During the signaling period, innovators developed new technologies. Indeed, anticipation of regulations is the main motive for environmental innovators. More than two-third of innovators request subsidies for this purpose (Arentsen and Hofman, 1996). Although the development of environmental technologies also takes time, it is sometimes more than ten years, the development period is usually much shorter than the signaling period. So pressure for high R&D expenditures to shorten the lead time to market in view of competition, as found on private markets, is hardly relevant in the environmental policy making.
b. Preparation period
Policy preparation starts as the signal is incorporated by involved administration, business branches and politics. Preparation usually starts with an inventory of emission sources and demonstration of technologies that can be used to tackle the signaled problem. It means preparation of emissions standards is based on the performance of technologies available at that moment. The successful R&D investment in the signaling period enables demonstration of the new technology for approval among the best available technologies (BAT). The technologies must be available at the start of policy preparation because the costs and effects must be demonstrated at some emission sources and performance must be verified by some experts. Thereafter, the BAT are approved in expert groups with representatives of authorities, industries and scientific institutions. The administrative BAT qualification takes three to five years, but disputes about costs and effects cause delays (Sørup, 2000). The political approval process that follows incorporates opinions and interests at various departments, lobbies of interest groups, budget constraints for policy making, industries' investment from the past, diversity of emission sources, the costs of emissions reduction and so on. This has taken four to eight years in the Netherlands (Klink, et al., 1991:70–80; Van der Straaten, 1994:131–145). It is even longer in the EU. For example, the Convention on Long

Range Transboundary Air Pollution was agreed between EU member countries in 1979, signed by the European Commission four years later and announced as the Directive on Emissions to Air from Large Combustion Plants six years after that in 1988. Generally, the preparation of emission standards takes about 7 to 12 years. The period can be shortened if politicians sense high urgency, but it is often longer because some interests are able to negotiate delays. The process of policy preparation ends with the announcement of a directive that provides a legal basis for enforcement of emission standards. The covenants are prepared in a similar way, but take less time because they are agreements between interests that can be enforced without political debates. However, a covenant is often a precursor of the preparation of emission standards. The preparation of economic instruments takes a longer time than preparation of emission standards because the economic instruments are usually opposed by various interests and because they must be incorporated in the countries' fiscal policy. This entails co-ordination between departments. To avoid the latter, one finds private arrangements about funds based on fees, for example, for packaging waste, batteries and cars. Subsidies for technology development are usually well accepted by interests, so the preparation phase is shorter. More debated are subsidies for investments and for emission reduction because the volume of subsidies is difficult to control.

c. Enforcement period

Enforcement of emissions standards after policy preparation entails implementation of environmental technologies at emission sources until a sufficient number of sources addressed during policy preparation is covered. The implementation means installation of technologies that meet the requirements at all major emission sources which is usually based on a mandatory emission reduction at every large source. During enforcement, innovators can sell the technologies prescribed in permits. The sales of innovations remain uncertain during enforcement because it is often not the best performing technologies that are enforced, but rather the technologies that are acceptable for authorities and companies. Legal procedure and administrative capacity hamper dissemination of new technologies. It holds for the large emission sources, for example, the EU directives on emissions from domestic waste incinerators (89/429) and SO_2 and NO_x emission from large combustion plants (88/609) have not been fully implemented even 12 years after the start (Glachant, 2000). The domestic waste incinerators are large scale facilities that are easy to address. The enforcement of emission standards in the small and medium size companies is more laborious and even slower is implementation at small emission sources because of limitations in administrative capacity. A major cause of slow diffusion is the practice of revising the license only when the firm undertakes a major production change. This implies delays in sales of environmental innovations for many years because of the economic lifetime of installations that is usually ten to fifteen years (Klink et al., 1991:80–81, Glachant, 2000:9; Haq et al., 2001:130). Administrative restructuring also cause delays. For example, several EU member states have postponed enforcement of product-oriented environmental regulation by more than six years during policy preparation by the European Commission (Rubik and Scholl, 2002). The period of enforcement takes 10 to 15 years for the large installations and even longer in case of small emission sources. The enforcement period can be even longer because of exemptions, like complex administrative changes. For example, the EU's Water Framework Directive approved in 2000 projected an enforcement period of 15 years, but pressure to postpone it emerged shortly after approval. Companies are also permitted to postpone implementation because of other priorities. As a result, the sales of technologies are initially very low and are only growing gradually. In some cases, regulation is enforced rapidly to solve an urgent problem or to create a competitive advantage for domestic industries. This causes a high rate of diffusion of new technologies. The latter type of enforcement, so-called "strategic marketing", is widely disputed as non-tariff barriers in trade. It has been issued with regard to regulations on pesticide residues in foods, toxic dyes in clothes, equipment for oil losses on ships and so on. Using economic instruments, implementation is faster because polluters have financial incentives to reduce pollution. Covenants can be implemented in line with the agreement, but if firms are obstructive, implementation is delayed (Wit et al., 1999; Hofman and Schrama, 1999).

d. Evaluation

The policy cycle ends with evaluation of the results and eventual preparation of new regulations. Growth of production, studies on the impacts of pollution and failures to realize emissions reduction are major reasons for the renewal regulations. Only a few environmental problems covered by policy preparation in the 1970s passed the whole policy cycle including the evaluation and renewal. For example, it can be argued that policy preparation on emissions of biodegradable matter to water at the beginning of the 1970s was revised during the late 1990s with the EU Water Directive as the result. This is done more than 25 years later. Another example is the EU directive on catalytic converters for cars. This was announced in 1970 after several years of preparation (70/220EC) and adapted in 1993 (Haq, et al.: 131).

8.3.2 Innovation Cycle

The decisions of innovators who try to anticipate new environmental regulations can be related to the policy cycle.

e. Technology development period

The innovators who succeed in developing and demonstrating the new technology before policy preparation may be able to influence policy makers' choice of standard in their favor. Therefore, the technology development period starts somewhere around half-way after the signaling period and ends in the early years of the preparation period. In deciding to invest in R&DD, the innovator faces major uncertainties: Will the R&DD investment result in a technically feasible cleaner technology? If so, will the regulator qualify the innovation as a BAT? How fast will the implementation start enabling sales of the developed technologies? The technology development period and components of uncertainty address all policy instruments. In the case of economic instruments and covenants the emission sources have the flexibility to use all technological options, not only preselected BAT.

f. Waiting time

After the demonstration, innovators must wait four to eight years and sometimes longer during the preparation period before the environmental demands and instruments are approved by the authorities and enforcement can start entailing implementation at the sources. For example, the technologies to reduce sulfur from smoke stacks in the oil industry were tested in the mid-1960s but took effect as mandatory requirements more than 10 years later (van Driel and Krozer, 1987). Another example is the technologies to reduce sulfur emissions in the Finnish pulp and paper industries that were developed 10 years before being set in national guidelines in 1987 (Kivimaa and Mickwitz, 2004). Some innovators that anticipate regulations in leading countries have to wait about a decade for international harmonization of environmental regulations, as shown with the low NO_x engine on ships (Hyvättinen and Hildén, 2004). During the waiting period, innovators can customize their technology and promote it, but they cannot sell it because customers can and will wait until the regulation is enforced. The waiting period between the demonstration of the new technology and the political decision to enforce an emissions standard is a minimum of six to eight years. The waiting period can be shortened by publicity and other pressures, but innovators must often wait even longer before policy makers start with enforcement and new technology can be sold. The waiting time in the case of economic instruments can be longer because of resistance by polluting industries. The waiting period involved in covenants is shorter because they are easier to prepare than formal regulations, but it is still about four years or more.

g. Sales period

The new environmental technology can be sold during the implementation period, which is 10 to 20 years, as shown under heading (c). The innovators' interest is a short implementation period that is to sell as many innovations in the early phase of implementation period as possible because it enlarges the present value of sales. This can be supported by launching innovations with policy certificates and labels, as well as by economic instruments because they encourage fast reduction of costly emissions. The present value of sales is low if innovations are postponed, for example, because vested interests have a strong lobby or voluntary emissions reduction in covenants is obstructed.

h. Maturation period

Waiting time (e) and sales period (f) determine together how much time passes between the end of the R&DD period with demonstration of a new technology and the moment to cash in on the revenues from the innovation. As an index for the length of the sales period, which takes into account the distribution of sales over the period, we shall use the median of sales. That is, the number of years it takes to realize half the total potential sales revenue during the implementation period. The median of sales is usual in business (Ganguly, 1999). The sum of the waiting time (during preparation) and the time to achieve the median of sales is labeled the maturation period. The time between the innovators' decision to develop a new technology and sales is very long. Environmental innovators must wait a minimum of six to eight years during policy preparation and accept that enforcement can take a minimum of 10 years. So, the lead time between the investment in R&D and saturation of the sales is 16 to 20 years.

In Scheme 8.1 the phases in the Policy cycle and in the Innovation cycle are shown in relation to each other.

Scheme 8.1. Phases in environmental policy making, innovation and dissemination in years

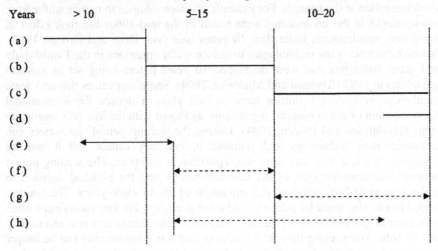

Years	> 10	5–15	10–20
(a)			
(b)			
(c)			
(d)			
(e)			
(f)			
(g)			
(h)			

The policy cycle

    a.   The signaling period

    b.   The preparation period

    c.   The implementation period

    d.   The evaluation period

The innovations cycle

    e.   The technology development period

    f.   The waiting period

    g.   The sales period

    h.   The maturation period

8.4 Uncertainties for the Innovator

During the policy cycle, innovators are uncertain about the possibility of technology sales. This is because the choice of technologies by policy makers during policy preparation and by companies and authorities during enforcement often does not reflect technological performance (such as low costs and high effects), but rather considers vested interests that are difficult for innovators to foresee. It is observed that considerations of vested interests usually cause higher implementation costs in comparison to the lowest-cost solutions; so-called X-inefficiencies (Rose-Ackerman, 1983; Nentjes, 1988). Thus, sales of environmental innovations during enforcement are less certain than can be expected on the basis of demonstrated performance during policy preparation.

Many factors cause X-inefficiencies in environmental policy. An important factor is differences in the tasks and interests of public and private organizations. Preparation of environmental demands is delegated to supranational authorities like the European Commission, but national authorities also play an important role in environmental issues. These, in turn are divided between various departments such as economic affairs for energy, water affairs for effluents, agriculture for space use and environmental affairs for emissions. Enforcement is usually delegated to local and regional authorities, such as municipalities and provinces, to specialized sector departments such as forestry and water boards, as well as to semi-public organizations, such as foundations to raise and manage fees and public utilities for waste management. In addition to regulation of standards that must be implemented because they can be legally enforced by law, there are also demands that lack a legal status. Implementation also depends on several market interests that can be in opposition. These issues cause uncertainty about the sales of environmental innovations.

We present two cases that illustrate the uncertainties about sales caused by authorities' different interests and differences in market interests. These address the policy on heavy-metal emissions released into water and VOC emissions from paints to air. The cases show that the preparation and enforcement of regulations is uncertain for the innovator because it is unclear if effective and efficient technologies are going to be implemented.

A case is presented on divided interests in policy making. An example of the interests involved in environmental policy that cause imperfection in use of environmental technologies can be found in the area of emissions of heavy metals released into water. This kind of emission is a major threat to public health and nature because heavy metals such as cadmium, mercury, chromium, nickel, zinc and lead are very toxic. The case is based on studies on the effects of emissions reduction of heavy metals by economic instruments on diffusion of environmental technology. The studies were done in the mid-1980s in the Netherlands (Klink et al., 1991; Nentjes and De Vries, 1990).

There are many sources of heavy metals. In households, the main sources are found in the construction of houses such as zinc gutter and lead pipes, as well as residues from paints and inks. Rinse from roads is the main non-point source. In manufacturing, heavy metals are used for anti-corrosives of iron and steel, in paints, for tanning leather, printing and catalysts. Heavy metals get rinsed out into sewage. They do not break down but only disperse. In the water-treatment plants located at the end of the sewage pipes, between 60% and 80% of the heavy metals are flocked down into sludge, the remaining part is dissolved into water. The sludge is often used as a fertilizer in agriculture and causes contamination of the food cycle. Another part is treated as solid waste, thus emitted to air through waste incinerators and comes back on the ground with rainfall, or it is directly landfilled. In view of the complex cycle of heavy metals, several authorities are involved in the prevention of contamination with heavy metals. The emission to water is regulated by the Ministry of Water Affairs and the Ministry of Environment; the latter department sets the emissions standards and enforces the standards in waste management, the former department sets the standards on water that are enforced by the regional water authorities. The prevention of soil contamination is regulated by the Ministry of Agriculture. The main policy instrument is the emissions standard that defines the maximum concentration of heavy metals in a medium such as effluent or exhaust gas. The standard for water is set in such a way that an available technology (neutralization) can be used. There is also a charge to finance public water treatment. In most EU countries, the charge is placed on heavy metals in water, not in sludge. Most polluters use the neutralization to flock down the emission from water to sludge. Hence, heavy metals move from water to waste because it is the cheapest way to reduce emissions in water. It is possible to prevent the dispersion of heavy metals by recycling in process, which entails separation of the heavy metals from water and reuse in process or sales to other customers. There are recycling technologies, but they are too expensive except for some expensive metals like nickel or silver, as long as sludge disposal is cheap. A charge on heavy metals in sludge and in water can invoke recycling. Studies on the effects of charges on heavy metals in water and sludge on the recycling technologies indicate that a moderately high charge can invoke recycling. The costs are initially high, because the companies have to pay for the discharge to water and sludge, but as neutralization is gradually substituted by recycling technologies, the costs go down substantially. So while there are the costs of change-over, recycling technologies are net beneficial in comparison to the use of neutralization technology and a charge on heavy metals released into water and sludge. Recycling technologies are not implemented because of strong resistance to the charge on heavy metals and the division of tasks between water authorities and the waste authority.

This case is on divided interests in market chains. Many emissions are caused by the use of products. Product-related emissions can only be reduced by product changes, but differences in market interests in the chain impede the changes. An example of impediments is the substitution of VOC-rich by VOC-low paints. The case is based on studies done in the mid-1980s and mid-1990s (Kremers, et al., 1991; Derksen and Krozer, 1996). Substitution of VOC-rich by VOC-low paints is agreed within the framework of the covenant on VOC, embracing many emission sources (KWS-2000). The agreement is set between authorities and companies that produce paints, whereas the users of paints are not included in the agreement. Six years later it was found that the sale of VOC-low paints and other VOC-low products is slower than was expected at the time of agreement. This is because many professional painters perceive VOC-low paints to be of lower quality and because their use is more costly. The higher costs of VOC-low paint are not so much caused by higher production costs, but rather by mark ups in wholesale and by professional painters. The mark up has to do with the scale of sales and experience in application. Environmental authorities discussed enforcement of economic instruments that would level out the additional costs. A low fee was proposed on the VOC-rich paint in production and import or a slightly higher fee in distribution with the restitution of the fee for the VOC-low paint. The administration and control costs of the system are low, only 3% to 8% of the total cost of emissions reduction; the distribution fee is particularly easy to administer. The proposal for the fee was refused because the producing companies expected it to have a negative effect on paint sales, whereas professional painters demanded lower prices and better quality. The fee in distribution was opposed by retailers that were not included in the covenant and because the authorities expected difficulties with legislation if the companies did not co-operate on a voluntary basis. No solution was found after five years of negotiations between the paint producers, the users of paint and the authorities. Compliance with the covenant is hindered by the different market interests that are acknowledged many years after signing the agreement. Product-related VOC emissions have been barely reduced, whereas process-related emissions reduction progresses because it has been enforced in permits in the case of non-compliance with the voluntary agreements. The covenant did not include instruments for product-related emissions reduction.

8.5 Model for Environmental Innovations

The waiting period and uncertainty caused by imperfections in policy preparation and enforcement are relevant factors for environmental innovations. The model that enables assessment of the main factors that influence profitability of investments in environmental R&D is elaborated below.

We take the position of innovators who are aware of possible sales during enforcement and the necessary investment in technology development before policy preparation. In view of the signals about a new environmental issue, an innovator expects stricter environmental demands and decides to invest in R&D to anticipate policy making. The innovator develops technology in advance of policy preparation. The innovator is able to demonstrate the excellence of the technology to the authorities on time and to lobby for enforcement at many emission sources. So the innovators take into consideration the waiting period during policy preparation and the imperfections during the enforcement of environmental demands. The innovator decides rationally with the help of the Net Present Value (NPV). Future sales revenues of the new technology are discounted at an interest rate that reflects the uncertainties in policy preparation and in enforcement, that is, the more uncertainty about sales the higher is the interest rate. The question is: What are the factors that determine the profitability of the investment in innovations? The model is presented in Figure 8.2. Aside from the sales volume and investment, the factors are:

a) The waiting period, that is the period between the end of a successful demonstration of the innovation and the start of enforcement;

b) The enforcement period, that is the period between the start of enforcement and the final sales of the technology, because all relevant emission sources have complied with the demand;

The uncertainty about the consequences of policy preparation and enforcement for the sales of innovation in the future is reflected in the interest rate for discounting the revenues.

R&D investment and sales revenues (present value)

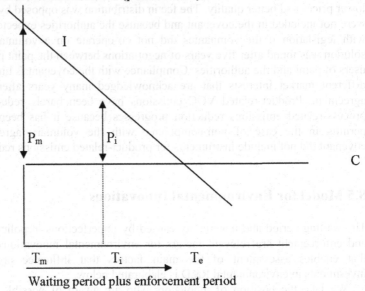

Waiting period plus enforcement period

Figure 8.2. Effects of the waiting period and enforcement on expected revenues and profit

Vertically, the investment in innovation C and the revenues of sales I are shown. Horizontally, the years between the realization and sales of the innovation are presented that are the waiting period plus the enforcement period. The discounted sales must cover the investment. The investment in R&DD is done before the waiting time, so it is presented as a horizontal line independent of the duration of policy making. Strict environmental demand can require a higher investment because more advanced technologies must be developed, i.e. the line C can shift up or down depending on the demand. The present value of sales, presented as a decreasing function of time, is influenced by the interest rate, i.e. a steeper line, I, at a higher interest rate. If there is no waiting period, the sales start directly after the announcement of the demands, then the innovators can expect revenues at the moment Tm, and realize profit that equals Pm. This situation reflects the interest rate on private markets when demanders have interest to buy and use the innovation superior to the available technology. In the environmental policy, however, innovators face the waiting time and the enforcement period. If the innovators must wait until the moment Ti because of laborious preparation and enforcement, then they can realize profit Pi and if the waiting period is longer until Te then there is net loss because the present value of the revenues is lower than the investment. Thus, a longer waiting period or higher interest rates cause less profitable investment. If the innovators expect a long waiting period or uncertain enforcement then they withdraw from investments in technology development. The model also helps to analyze the effects of policy instruments on innovations:

- A shorter waiting period (e.g. covenant) or enforcement period (e.g. tradeable emissions) shifts the revenues function to the left and increases profitability of the investment in innovation;
- Uncertain preparation (e.g. emission charge) causes a higher interest rate and steeper sales line, hence a smaller profit in a discounting period;
- Lower investments in R&D (e.g. subsidy) shift the cost line downward from C_0 to C_1 and increase profitability of the investment.

The model is used to simulate the effects of policy instruments on profitability of investments in innovations taking into consideration the waiting period during policy preparation and different types of enforcement. The indicator of profitability is the innovation rent, that is the expected profit of innovators plus the cost savings of the users due to sales of the new technology. For convenience of the accounts, we assume that the innovators accrue all the savings of users, i.e. sell on the margin of cost saving. It is formally:

Innovation rent

$$s_t = \frac{(v_t - o_t)}{r_i^t} \tag{8.1}$$

For $t = t_w + t_e$ and $r \geq 1; r_1 = 1; r_i > 1$ (8.2)

In relation to interest

$$S = \frac{\sum_{t=1}^{n} s_t}{r_i^t} - I \tag{8.3}$$

Index of surplus

$$Index = \frac{S(r_i)}{S(r_1)} \cdot 100 \tag{8.4}$$

v_t is the revenue from innovation sales; o_t are operational costs, i.e. the production costs of the new technology; I is the investment; S is the surplus of innovation rents s_t, r is the interest, t_w is the waiting time, t_e is the enforcement time.

To illustrate the effects of policy instruments on the innovation rent, we draft scenarios for policy preparation and enforcement. The following numerical values are used. The total market volume is 1 500 units in all scenarios ($V = 1\,500$). The operational costs are assumed to be 50% of sales ($o_t = 0.5 * v_t$). The present value of the investment in R&DD is 100 ($I = 100$). The interest, r, is: 100%, 105%, 110%, and 115%. The enforcement period, t_e, is 15 years and the waiting time, t_w, is 0 years, 4 years to 8 years, so the lead time of innovation and diffusion, t, is 15 years for 0 years of waiting time, 19 years for 4 years of waiting time and 23 years for 8 years of waiting time. Three types of enforcement are simulated. Reference enforcement that illustrates application of emission standards, with sales of 100 units per year over 15 years (100 units * 15 years = 1 500 units). Slow enforcement (low diffusion rate) that illustrates a covenant or a subsidy, starting with the sales of 2 units in the first year, building to 500 units in the last year and linear interpolation in between. Fast enforcement (high diffusion rate) that illustrates an emission charge or tradeable permit that is a reversed version of slow enforcement:

sales of 500 units in the first year and 2 units in the last year with linear interpolation in between.

The calculations can be found in Appendix D. Table 8.1 presents the results of the simulations. The results are indexed. The index is based on the surplus of the innovation rent during the enforcement period divided by the reference (100 units) that is the surplus without the waiting period and without discounting. The results show the surplus without investment in innovations due to the subsidy for technology development and with the investment of 150 units.

Table 8.1. Simulation of the effects of interest and waiting time on the innovation rents

	Gradual enforcement			Slow enforcement			Fast enforcement		
	0 years	4 years	8 years	0 years	4 years	8 years	0 years	4 years	8 years
No investment in innovations that is a 100% subsidy for technology development									
100%	100	100	100	100	100	100	100	100	100
105%	70	57	47	53	44	36	87	72	59
110%	54	35	24	30	20	14	77	53	36
115%	40	22	13	17	10	6	69	39	23
Investment in innovation of 150 units that is 10% of the total market volume									
100%	100	100	100	100	100	100	100	100	100
105%	62	46	34	42	30	20	84	64	49
110%	39	18	5	12	0	−8	71	41	20
115%	25	3	−9	−3	−13	−18	61	24	3

In this way, the consequences of the waiting period and the enforcement period on the profitability can be indicated for various instruments. The results of subsidized investments are between parentheses.

If we assume a 110% interest for an investment in innovation, eight years of waiting time and gradual enforcement, in the case of emission standards then the innovation rent after 15 years of enforcement is only 5% of the reference (24% with the subsidy). The innovation rent is 34% of the reference in a case having a 105% interest. A higher uncertainty about enforcement of emission standards strongly influences the profitability of R&D investment, hence an innovation spurt. With a 110% interest, only four years waiting time but slow enforcement, as can be the case with covenants, the surplus is 0% of the reference (20% with the subsidy). However, if compliance with the covenant is assured so the interest is only 105%, for example by liability for non-compliance, then 30% of the reference value can be reached despite slow enforcement. So, a covenant with liability can invoke innovations. At a 110% interest, eight years waiting time and fast enforcement, as can happen with economic instruments; the surplus is 20% of the reference value (36% with the subsidy). However, uncertainty about preparation and enforcement of the economic instruments, hence a 115% interest, can reduce the surplus down to only 3% of the reference. It is even negative in cases with a waiting time longer than eight years.

The results indicate that there is no "best" policy instrument from the innovators' point of view. Much depends on the balance between uncertainty and duration of policy preparation and enforcement. In general, uncertain policy making and a waiting period longer than eight years is usually unprofitable for innovations, albeit the waiting time during policy preparation can be compensated by fast enforcement. Thus, in comparison with emissions standards, one can argue that fast enforcement of policies due to economic instruments can counteract the negative effects of the waiting time on innovations' profitability, but strong resistance by polluters causes uncertain enforcement. The present covenants have a detrimental effect on the profitability of innovations, but this is not always the case. A short waiting period in preparation of covenants can be attractive for innovations despite slow enforcement and a smaller market volume in comparison with emission standards, if the agreements include liability for non-compliance. In addition, note that the waiting period and uncertainty are interlinked, so a shorter waiting time reduces uncertainty. In addition, it is simulated whether subsidies contribute to the innovation rent. The first type of subsidy is meant to support R&D, as mentioned above. The second one supports production of new technology, for example, the financing of manufacturing facilities. The simulations show that both types of subsidies have little positive effect on the innovation rent.

Based on the model, we can discuss implications of uncertainty and duration in policy making for policy instruments. Direct regulation by emissions standards will be used as the reference case. The analysis suggests that the positive effects of environmental policy on innovations largely depend on the possibilities to shorten the waiting period, to speed up implementation and reduce uncertainty about the regulator's acceptance of new, cleaner technologies. The policy instruments are compared in view of these three criteria in Table 8.2.

Table 8.2. Comparison of policy instruments in view of innovation theory

Policy instruments	Waiting period	Implementation	Diffusion rate
Reference: emission standards	4–6 years	10–20 years	Uncertain BAT
Subsidy technology development(*)	Similar: + / −	Similar: + / −	Higher: +
Subsidy procurement(*)	Shorter: +	Faster: +	Lower: −
Subsidy for emissions reduction	Longer: + / −	Faster: + / −	Higher: +
Emission charges (regulatory)	Longer: −	Faster: +	Higher: +
Tradeable emission rights	Longer: −	Faster: +	Higher: +
Covenant	Shorter: +	Slower: −	Higher: +
(*) in combination with the emission standard			

Preparation of the emission standards usually takes about 8 to 12 years, which implies that the waiting period after demonstration of a new technology and listing it as a BAT takes four to six years. The implementation is slow. It takes 10 to 20 years of diffusion before market saturation. Uncertainty about implementation of the new technology is high because it depends on administrative capacities.

Granting a subsidy in the early stage of policy making makes uncertainties more bearable because it reduces the innovator's investment expenditure on RD&D and the innovator's perception of uncertainty. A subsidy on the procurement of innovative technology speeds up implementation but has little positive effect on profitability. This is because the subsidy is granted in the sales period and discounting depresses its net present value. It is also timeconsuming and laborious to compound a list of subsidized technologies that reflects all relevant innovations. As a result, the procurement of past technologies is subsidized instead of innovations. A subsidy for emissions reduction needs more preparation time because the income for subsidizing must be generated. The implementation can be faster and diffusion of innovations can be enhanced but there are limits in granting.

The switch from direct regulations to market-based regulation, for example by regulatory emission charges, provides strong incentives for innovations. Although preparation time is usually longer because of opposition to emission charges, the implementation is faster because emission sources alone can decide about installation of suitable technologies and because sales are concentrated in the early years of implementation. This is due to the incentive for firms that pay charges or permit prices to save on expenditures for residual pollution.

The duration of preparation and the speed of implementation of tradeable emission rights are similar to emission charges. This means a long waiting period but fast implementation. Diffusion of innovations during implementation can even be faster under the regime of tradeable emission rights than under the emission charges because cleaner technologies enable an emission source to reduce emissions and sell the rights, thus generating income.

Covenants need a shorter preparation time than direct regulation. When covenants are in preparation, the innovator's R&DD decision is not basically different from the steps under direct regulation. The implementation, however, is slower because it is essentially voluntary with enforcement by direct regulation only in cases of non-compliance with the agreement. The demonstration of an innovation is still necessary to provide the relevant information but the major advantage of the covenant in comparison to the emission standard is that sources have the freedom to adopt the control options that suit their specific situation, which encourages the introduction of innovations.

8.6 Conclusion

The question was how to foster environmental innovations. The view that technology-forcing demands can be imposed by policy makers to trigger innovations is disputable because policy makers aim to be assured that companies can comply with the demands at reasonable costs. It is shown that the perspectives of policy makers and innovators differ. Instrument theory takes the view of policy makers. It suggests that economic instruments, like charges and tradeable permits are effective and efficient because polluting companies have to pay for residual emissions. Based on innovation theory, it can be argued that innovators must decide to invest despite uncertain future sales. For this reason, reducing uncertainty about enforcement of environmental demands is the main factor contributing to environmental innovations.

In practice, uncertainty about sales of environmental innovations is the decisive factor. It takes 20 or more years to prepare and enforce a stricter environmental standard at the most relevant emission sources. The innovator must invest to develop and demonstrate a new technology, then wait six to eight years until the standard is set and enforcement starts. During enforcement, the innovations can be sold. Innovators are uncertain if, when and how the standards are going to be approved and enforced by politicians. The lead time in policy making with uncertain outcomes is a major reason why high investments in environmental innovations are risky and unattractive.

The argument in the instrument theory that economic instruments provide a stronger incentive to innovate in comparison to emission standards is valid under additional conditions. The advantage of economic instruments is that a larger market volume for sales is created because polluters pay for residual emissions. In addition, economic instruments entail fast enforcement, because polluters alone decide about procurement based on their calculations, whereas enforcement of emission standards entails time-consuming negotiations between authorities and polluters. However, the waiting period of economic instruments during policy preparation can be very long because polluters resist the instruments. So, environmental policy must be strong enough to impose demands with economic instruments. If policy makers opt for covenants, for example, because they do not have a strong position with respect to other interests, then less market volume can be expected in comparison with emission standards and economic instruments, as well as slower enforcement, but the waiting time during policy preparation is short. So the loss of innovation rents by time-consuming enforcement can be somewhat compensated by fast preparation. Simulation of the profitability indicated by surplus of innovation rents suggest that more than eight years' waiting period and uncertain enforcement discourages investments in innovations. There is no perfect instrument but policy makers can create favorable conditions. Favorable conditions for innovations are a short preparation period of environmental demands with clear demands, reasonable assurance of enforcement and freedom for companies to act. Instruments that foster innovations shorten the waiting period, accelerate enforcement and reduce uncertainty because these three factors determine the profitability of investment in technology development.

9

Does Self-regulation Work?

The alternative for environmental policy making is negotiation about emissions reduction between various interests. The main interests at stake are social organisations such as environmental groups that represent the interests of harmed people, polluting companies that use environmental technologies and innovators who develop and sell new ones. We consider a situation in which policy makers do not intervene directly but instead provide a legal framework for negotiations. The question is: What conditions are needed for negotiations that can reduce emissions at low cost?

9.1 Self-regulation in Practice

In order to negotiate on emissions reduction, environmental groups must represent the interests of those who are willing to press for liability among polluting companies or to pay for improvements in environmental qualities. These kinds of interests, which we call stakeholders' demands, push companies to improve their performance. Stakeholders' demands undisputably emerge, although the effects of the demands on environmental innovation are limited, so far.

Stakeholders' demands make companies liable for unsafe and hazardous operations. Particularly large accidents have attracted attention and legal action, such as the chemical plant explosion in Bhopal, India and the oil spill caused by the tanker, Exxon Valdez, in North America. Intense media coverage sensitized companies to possible liability, raising the perception that liability for environmental damage can have major economic consequences for businesses. In reality, however, the liability related to such accidents has minor economic impact because such catastrophes fortunately occur rarely in comparison to daily operations. The annual costs of liability are low compared to many other cost factors in companies' operations. Minor accidents are often insured and the insurance fee is usually only a fraction of the total production costs. Exceptions are extremely hazardous processes such as nuclear power plants or hazardous waste incineration. Some damages are not insured such as soil pollution. Liability for such damages can be a burden to a company. On a statistical basis, however, the annual costs of liability are low in the EU. The annual cost of insurance for damage caused by pollution is only a small fraction of the annual total costs of emissions reduction. For example, in the Netherlands, a country that has been a frontrunner in introducing legal liability for environmental damage in Europe, the annual costs of insurance for environmental damage for all medium and small enterprises was about € 1.45 million during the 1990s. This is only about 0.5 per cent of the total pollution control costs carried by small and medium-sized enterprises. Unfortunately, more recent data are not available (Veering, 1993). Regarding such low insurance costs, environmental liability in the EU cannot provide substantial incentives for environmental innovations.

Companies' credibility is also a part of negotiations. A company's social responsibility plays a role in negotiations between suppliers and customers because it is a yardstick for overall performance. That is, a wasteful company is usually perceived as unreliable. Many companies have implemented management quality assurance (such as ISO 9 000) and some of them have gone beyond it, acquiring quality assurances for social management (SA 7 000) and for environmental management (ISO 14 000 or EMAS in the EU). The latter can be attained at a cost that is only a few percentage points above the costs of management quality assurance. However, despite the low cost, companies' response to environmental management quality assurance in the EU is low. The number of companies that acquired the EMAS certificate by 2002 in the EU slightly surpassed 1.6% of the manufacturing companies. Growth of certifications was saturated thereafter. The certificated companies are mainly large manufacturers aiming to legitimize their efforts with regard to policy makers. The effects of certification on environmental innovations are unclear (Heinelt, et al., 2003). The number of companies in the EU with ISO 14 000 certificate is also low. It adds a similar percentage to the environmentally certificated companies in the EU (Bridgen, 2003). The effect of such certificates on innovations is unclear. The companies' credibility has a value in negotiations with authorities about permits but it is insufficient to obtain a license to produce (Kaarrer-Rueedi, 1996). Financial institutions also monitor companies' environmental performance regarding assessments that the companies with high social and environmental rankings tend to be less sensitive to economic fluctuations than other companies, albeit slightly less profitable but the effects on innovation are uncertain (Konar and Cohen, 1997; Skillius and Wennberg, 1998). The demand for environmentally certified products, discussed in Chapter 6, seem to provide more opportunities to innovate. Despite the assessment, we are not aware of studies on the effects of high credibility on innovativeness.

The policy, based on stakeholders' negotiations on environmental qualities, was introduced in the 1980s under the term "self-regulation" (Winsemius, 1986). Various instruments for self-regulation have been envisaged and incorporated in policy making such as liability for toxic compounds in food, liability for soil pollution, producer responsibility for chemical waste, benchmarking of production, product certificates for health and environment, tradeable emission rights for pollution, and so on. Several internal company instruments are enforced, like environmental accounting systems, information and quality assurance in trade, producers' responsibility for products and others. Public control mechanisms have also been developed, including publicity and tribunals, public accountability by companies and others. Direct negotiations between stakeholders, however, are rare. In some cases, environmental groups negotiate with companies about emissions reduction. Sometimes they do this in a coalition with other businesses involved in the life cycle of products, for example with retail businesses on preventing hazardous compounds in foods or with trade unions about banning imports of unqualified garments. Some environmental groups acquire farmland and make agreements with farmers on so-called ecological production. Again, the effects of self-regulation on environmental innovations remain unclear.

Based on scarce experiences, it can be hypothesized that self-regulation works if the rules of the game are enforced, controlled, and if the negotiations are pursued for tangible reasons, such as regional planning, nature management, local nuisance, consumer products and waste prevention. One needs to examine when self-regulation involves environmental innovations. This is important because the low costs of environmental policy broaden public support for environmental quality and strengthen the position put forward by those demanding a high-quality environment.

9.2 Self-regulation in Theory

The negotiations between stakeholders on environmental quality are at the core of mainstream economic theory. In this theory, the allocation of property rights for social goods, such as the environment, involves negotiations between private interests. Negotiations, in the theory, lead to optimal distribution of the goods, independent of the initial allocation of rights. This occurs if the negotiations take no time and do not cause transaction costs, which implies that there is perfect knowledge about whose interests are involved, the solutions that are available and how the goods can be distributed between private interests, for example, by transferring ownership of land or water. In such idealized negotiations, a policy maker must define the rights of the interested parties, their possibilities to execute their rights and provide room for the negotiations. Having the right enables an interest to impose liability on those who cause harm and gain compensation (Coase, 1972). This is the essence of the so-called Coase theorem. The theorem is heavily disputed by different points of view. Only a few main arguments are briefly noted here because reviews of the dispute can be found in various excellent works.[10] One argument is that initial inequalities in power and expertise persist, which cause interest to pursue private gains at higher social costs, called free-riding. Another argument is the inability to privatise many public goods such as the climate or the ocean. The irreversibility of damages is also argued, such as damage to health. Still another one is uncertainty about the impacts of emissions regarding

---

[10] The mainstream perspective on self-regulation is presented in Pearce and Kerry Turner, 1990:70–83. The game theoretical approach is discussed in Weimann, 1990. The perspective of political theory can be found in De Beus, 1991. The negotiations process is analyzed in Kip Viscusi et al., 1995: 711–719. The discussion in view of various environmental problems is reviewed in Ostman, et al., 1997.

complex interactions that require laborious negotiations about emissions reduction, as in the case of climate change. The possibilities of self-regulation by stakeholder negotiations, therefore, are heavily debated.

Many argue that self-regulation is an effective instrument in environmental policy making. Some underpin the argument with management responses on inquiries about environmental qualities that underline the importance of stakeholders' demands on companies' image, marketing, motivation of staff, liability and on financial institutions (Bressers, 1995). It is also argued that the leading companies have internalized environmental interests within their organizations because they increasingly interact with various parts of society on a co-operative basis, which is instrumental for operations (Tomer, 1992; Schrama, 1987). It is also pinpointed that an increasing number of companies already have a profile as socially responsible organizations (Ianuzzi, 2002; Tomer and Sadler, 2007). Other scholars are more cautious about self-regulation. It is advocated as well, but this is generally done with caution about industries' responsiveness to stakeholders' demands. They argue that self-regulation should be considered an addition to other policy instruments combined in a mix (Cramer and Schot, 1993; Gunningham and Grabowsky, 1998; Noci and Verganti, 1999; Vickers and Cirdney-Hayes, 1999, Hofman, 2001). This argumentation, however, undermines an objective of self-regulation that is to reduce the regulatory inefficiencies mentioned in Chapter 8.

Advocates of self-regulation also considered the effects on environmental innovations. It has been argued that environmental demands are already so forceful that innovative companies can reduce emissions at lower costs than laggards, reduce their liability risk, harness credibility and they can sell solutions to other polluters. Networks between companies and environmental organizations would create an arrangement to develop environmental technologies and introduce innovations in specific social networks (market niches). It would reduce the risks of collision between various goals, the difference in company and user cultures, pervasive but unsuitable infrastructure, innovations' negative social and environmental side effects and so on (Verheul and Vergragt, 1995; Kemp et al., 1998). It is also pointed out that companies' R&D departments increasingly search for environmental innovations despite the higher complexity and costs of R&D. This is because management is aware of the need to integrate environmental issues in product development (Roome 1994, Clarke and Roome, 1995). The actual effects of stakeholders' demands on corporate R&D are also noted in a number of large companies (Cramer, 1997; Benn et al. 2007). Other studies, however, reveal that the effects of stakeholders' demands on corporate R&D are low despite growing awareness (Heinelt and Töller, 2003). These effects, if found, are mainly due to policy making. The observation that firms are reluctant to innovate unless they are pressed to take a risk on innovation has invoked much theorizing about risk-reduction strategies for environmental innovations through the networks. Several scholars have argued that social networks dedicated to environmental qualities would create market niches that enable the introduction of environmental innovations. Such dedicated networks would foster know-how, create sales opportunities and reduce investment risks (Kemp et al, 1998; Johansson, et al., 2004; Jorna, 2006). This view is advocated despite observations that building

coalitions between various interests to create a market niche is extremely difficult
and that companies must be able to cope with various often contradictory demands
(Szejnwald-Brown, et al. 2003). A study on companies' co-operation on
environmental innovations has shown that the determining factor for success is
trust between key partners in R&D projects. It is also observed that stakeholders'
interests are difficult to incorporate into a trusting partnership (Karl, et al. 2005).

Many doubt the possibilities of self-regulation. Firstly, there are different
interests in the chain. It is argued that companies in basic industries such as
agriculture, mining, refineries, metal, building materials and chemicals – usually
the largest polluters – must spend a lot to avoid liability, whereas companies on
consumer markets that cause less pollution can benefit from their products' good
image. Consumer-oriented companies that are sensitive to consumer demands can
influence suppliers, if environmental values are demanded in products. However,
the demands are ambiguous because consumers want product qualities at the
lowest possible price for private use, and at the same time, as citizens, want
products to have a high environmental profile, which raises prices (Dutilh, 1995).
The different interests of companies in the chain are confirmed by a study on
voluntary participation in the program on reduction of hazardous waste in the
United States. The program, called the 33/50 Program, started in 1988, aimed at a
33% reduction of such wastes by 1992 and a 50% decrease by 1995 in comparison
with the amount at the start. Due to the program's voluntary nature, it was expected
that companies participated randomly with respect to sectors, activity and size, but
actually the companies with large toxic inventories and those that operated on
consumer markets participated significantly more than others. Despite the
participation of large toxic inventories, the large polluters, the voluntary program's
effectiveness was low. The majority of the targeted reduction for 1992 had already
been achieved before 1988. During the entire program period 1988–1995,
emissions per unit were reduced by 50%, but total emissions remained at the 1988
level because of increased production (Arora and Cason, 1996). Secondly, high
transaction costs involved in self-regulations were pinpointed. Since information
about products' qualities is not freely available, the exchange of such information
requires transaction costs that increase exponentially as a function of additional
qualities in the negotiations. An exponential increase can be expected because
suppliers must spend more to promote the superiority of their products. They must
also do this as customers' capability to distinguish between the qualities decreases,
which in turn, requires even more promotion to convince the customers (Foss,
1996). The argument has some statistical support; for example, in the Netherlands
during the period 1980 to 1996 that coincided with the introduction of self-
regulation, the statistically measured costs of environmental management and the
costs of policy making, which approximate transaction costs, grew faster annually
on average (2.3% to 5.8%, respectively) than environmental investments and R&D
(2.1% to 1.9%, respectively).

There are also doubts about the effects of self-regulation on environmental
innovations as environmental issues play a limited role in companies' decisions
about innovations. A study on the role of environmental managers in large
chemical companies has shown that managers play a dominant role in negotiations
with authorities on regulations, but have a low profile in companies' decisions

about innovations. Environmental managers rarely co-ordinate the innovative process. They are not change agents in product development, nor do they function in the nexus of corporate R&D. Their main task is to select environmental technology that complies with regulations (Schot, 1991). Another study also revealed that companies have little expertise on environmental issues and they do not aim to generate it, as such expertise is not considered crucial for operations. Most companies rely on the expertise of external organizations, such as installation companies, intermediaries and advisors (Brezet, 1994). Experiences with co-operation between companies on ecodesign suggest that this co-operation is often unsuccessful, despite subsidies and the competitive advantages offered by co-operation. In addition to the barriers for innovations imposed by markets such as low market volume and low benefits for some co-operating companies, there are social impediments including the perception of an unequal position between partners, frustration felt by sector leaders about laggards and miscommunication in projects. More subsidies are advocated, but it is unclear whether these will provide better results (Georg et al., 1992). Studies on the selection of R&D projects in large companies confirm that environmental issues have a low ranking within the priorities of corporate R&D. Instead, the main criteria are competitor targets and the company's area of key competence (Chiesa et al., 1999). It is also argued that stakeholders' negotiations are pursued in a setting that is dominated by the vested interests of large companies that tend to postpone regulations, delaying innovations (Ashford, 2005).

The possibilities for environmental innovations using self-regulation are so controversial that theoretical and descriptive studies are insufficient. Analyses of experiments with self-regulation aiming at environmental innovations provide more insight into favorable conditions. Unfortunately, there are not many such experiments. An interesting one has been carried out in the Netherlands within the framework of the Packaging Covenant, which is analyzed as an instructive case study.

9.3 Experience with Stakeholder Negotiations

The issue of packaging waste is apparently an ideal possibility for stakeholder negotiations. Packaging is tangible for all consumers. Packaging is also a know-how intensive product, particularly with regard to product design. Many companies must deal with it and various alternatives have been discussed such as changing from plastic to biodegradable materials and using return systems. There are also opposing interests in the life cycle, notably the producers of packaging that earn income from the sale of additional packaging and the receivers of packaging who bear its costs and demand to reduce waste. Packaging materials are widely considered wasteful. It annoys many because it is a major source of litter.

All these interests had to be considered by the Ministry of Environmental Affairs and Ministry of Economic Affairs in the Netherlands before urging the packaging producers and packers, those industries that use packaging for products, to sign an agreement to reduce packaging waste. The agreement, for which preparation started in 1987, was signed by the Minister of the Environment and a foundation representing a few hundred companies in the packaging chain in 1992. The chain covers suppliers of packing materials, producers of packaging (including construction and printing), packers that use packaging, retailers and packaging recyclers. The Packaging Covenant specifies the targets for each type of packaging with direct regulation being used as a threat in the case of non-compliance. Its overall aim was to achieve 10% prevention and 60% recycling of packaging waste by the year 2000 (in comparison with 1990 levels). The Covenant preceded the EU's Packaging Directive that came into force in 1996. In this way, policy making in the Netherlands has provided a regulatory framework but refrained from direct interventions because implementation of the agreement is left to the companies. Through the foundation, companies created working groups on various product groups such as dairy, spreads, toiletries, chemicals, transport and so on. Implementation progress was monitored and controlled by a Board comprised of various stakeholders, including environmental groups that were represented by a consumer organization (the author was advisor to the Board and to several of the working groups).

The Packaging Covenant provides a suitable case to assess the effects of self-regulation on environmental innovations; the innovations in this case contribute to the realization of the targets. The advantages and disadvantages of the Packaging Covenant have been widely debated but these issues are left out of the study. The focus is on the process and the way stakeholder negotiations influenced environmental innovations in packaging chains. The study illustrates many positive adaptations due to the negotiations but also suggests that larger social benefits could be gained from innovations in theory, but they are not actually attained because of colliding interests in the chain. This is not necessarily because of the interests represented by consumer and environmental organizations but also because of colliding interests between companies in the chain (Doelman and Krozer, 1994; Krozer et al., 1995).

Some reduction in packaging can benefit companies in the chain. Although packaging is used for various functions involved in the delivery and sale of products, such as protection, attractive design, customer information and theft prevention, it is partly useless at the retail level and wasteful for consumers after product use, although these two bear its costs. On average, packaging constitutes 4% to 10% of the final product price at production. The highest percentage is for foods and toiletries. The annual costs of packaging in the Netherlands are about € 4 billion. Distribution adds to packaging costs on the market because it uses shelf space in shops, it requires handling with packaging materials and it causes waste. The share of packaging in consumer product prices increases to 15% and 25% for some products. Hence, packaging is a large cost factor and retailers and consumers have an interest in reducing packaging as long as it does not undermine a product's qualities. The costs bring a market incentive to audit the packaging functions, negotiate reduction of avoidable packaging and adapt packaging to reduce volume and weight. Environmental innovations that reduce packaging without functional losses in the chain, therefore, are beneficial to private interests and contribute to social welfare (a typical "win–win" situation).

The Covenant's socio-economic and environmental effects were assessed shortly after the agreement was signed. The assessment covered scenarios for costs in the chain, effects on labor and waste reduction for the period 1991 to 2000. The labor effects are not discussed here as these are not relevant in the context of negotiations about environmental innovations in packaging. The scenarios are based on international literature on packaging, interviews with experts and databases from the Netherlands Packaging Centre, an institute that supports packaging businesses. The assessment covers several groups of paper, board and plastic packaging. The other material groups, metals, glass and wood, hardly add to the overall results because of their small value and weight in total packaging in the Netherlands. The group paper and board covers about 55% of all packaging weight. The assessment of innovations includes corrugated board, massive board, cardboard, paper sacks and milk packs. This covers about 70% of the total packaging made from paper and board by value. The plastics group covers about 20% of all packaging by weight. Plastic packaging is very diverse. The assessment of innovations covers crates and boxes, bottles, sacks, foils and tubes. These constitute 38% of the total plastic packaging value. In total, the assessment of environmental innovations covers roughly about 40% of the total value of packaging. Four types of innovations are found:

- substitution of packaging materials such as substitution of corrugated board with plastic crates or substitution of paper sacks with plastic foil with weight reduction;
- reduction of the amount of materials used per packaging, such as lighter plastic bags (possibly with a deposit-refund), or lighter tubes and bottles for foods;
- reduction of packaging per product, including the elimination of cardboard around toothpaste tubes, or less frequent use of blister that is packaging with board on the back and plastic on the front;
- additional material and product reuse, such as substitution of milk packs with bottles using a return system.

Innovations are used to assess the possible effects policies and negotiations have on costs and waste in the chain. The steps in the chain cover: production of packaging materials, production of packaging, finishing packaging like printing, packing products, transport to wholesale and retail, collection and incineration of packaging and finally, recycling. The cost and waste streams are assessed by step in the chain because the effects on value and waste differ for each step. For example, prevention of packaging causes lower sales and turnover at production of packaging, but also causes savings at the retail level due to lower shelf coverage and less waste. This results in lower total costs in the chain. In the opposite way, return systems generate extra costs in retail and packing because packaging must be transported back and forth. However, there are lower weights and costs in collection, incineration and recycling of packaging waste.

Table 9.1 summarizes the scenarios of costs and waste in the Netherlands between 1991 and 2000. The results are the summarized costs and waste of the steps in the chain. They are indexed for the reference year 1990. Four scenarios are drafted. The reference scenario is based on the extrapolation of past trends, which show about a 3% annual growth in packaging weight. The goal of the Covenant cannot be attained, but suppliers of packaging gain a lot. The result of the extrapolation is more waste (+28%) and costs (+23%) in the chain in the year 2000 in comparison with 1990, mainly because of higher logistic and disposal costs. In the return scenario, the prime goal is recycling. In this scenario, policy making is the driving force because there is little to be gained from market interests. The costs and waste are assessed on the assumption that returning and recycling are enforced. A take-back policy increases costs (+10%) and reduces waste (–9%), which means that the recycling goal included in the Packaging Covenant cannot be reached as recycling causes heavier materials. The market scenario is mainly based on massive changes in packaging implemented by a few retailers in Germany and Switzerland. These have enabled a decrease in packaging use in retail and consumption. It is assumed that receivers of packaging (the retailers and consumers) are the driving forces behind it. Dissemination of new packaging can reduce costs by 7% and waste by 11% in comparison with the reference, but the recycling goal cannot be attained. The scenario shows net benefits for retailers and consumers due to less packaging, but losses in production. The innovations scenario is based on the market scenario. However, it is assumed that some packaging is substituted by new products envisaged in the literature and by experts. It also assumes that these products will get a better price, for example, due to the packed products' greater durability. A price increase of 3% a year is assumed for such innovative packaging. In the innovation scenario, the costs in 2000 are more or less the same as those in the reference year 1990, but it is possible to reduce waste up to 14% and to approach the recycling goal. The scenarios suggests that environmental innovations in packaging can generate monetary benefits in the chain and reduce waste, thereby approaching the aims of the Covenant. The reality, however, has been different, which is discussed below.

Table 9.1. Scenarios of annual packaging costs and weight (indices) in the case of Covenant Packaging for board, paper and plastic in the year 2000

Scenarios in the year 2000	Index consumers' costs (in € mln in 1990)	Index waste (1000 tonnes in 1990)
Basic year 1990	100 (4 295)	100 (1 463)
Extrapolation (supply-driven)	123	128
Take-back (policy-driven)	110	91
Market (demand-driven)	93	89
Innovations (quality-driven)	100	86

In the Covenant, it was agreed to conduct studies on the environmental impact of packaging with the focus on comparing one-way and returnable packaging. The companies added market-economic analyses to this agreement. The assessment is made per product group such as packaging for milk, sauces, spreads, bread, washing detergent, toiletries, and so on. This assessment work has been supervised by the working groups comprising representatives of businesses in the chain and consumer organisations. Government authorities were not represented and it was solely monitored by the Board. The working groups had to agree on the study's methodology and on the follow-up based on the assessment's results. The follow-up could be adaptation of the existing packaging, implementation of returnable packaging, or other innovations. The working groups provide an illustration on negotiations between interests regarding environmental issues.

The studies invoked negotiations on the assessment criteria, the reliability of the analyses, as well as deliberation on environmental versus economic advantages. The process has caused a delay of three years in comparison with the planned two years of studies. The studies have found that high costs are put on the packers (industries that use packaging for products). This finding is contrary to the assumption about a short time and low costs of the negotiation process under the theory on self-regulation. In reality, the negotiation process delays implementation and entails high transaction costs.

A broad consensus has been achieved that returnable packaging is often ineffective and inefficient in comparison to available, one-way packaging, if the one-way packaging is improved. This finding is contrary to environmental advocacy. Only returnable packaging for milk products is found to perform better from an environmental point of view, but less so from an economic perspective. Due to pressure from environmental groups and the supplier of an innovative, returnable packaging based on a plastic material called polycarbonate, a returnable plastic bottle with a deposit-return system has been introduced for milk products. It was resisted by the dairy industries, but the joint interest of environmental groups and the supplier of the plastic bottle, though not as a formal coalition, have brought about the innovation. This underpins the theoretical assumptions of self-regulation that the results of stakeholder negotiations can be independent of the initial distribution of rights.

The negotiations generated much expertise on the factors that influence costs and environmental effects in the chain. Based on this, various product adaptations have been implemented such as thinner tubes, lighter bottles, plastic crates, cardboard boxes, and so on. Many adaptations were available before the Covenant, but the negotiations triggered their implementation. Inquiries on the opinions of companies that participated in the working groups confirmed that the negotiations enlarged awareness on the environmental and economic issues involved in packaging. They also brought about adaptations in companies' processes. The positive overall effect of the adaptations has been that the growth of packaging waste has stopped and been reversed to waste reduction, although the reverse trend is not as strong as agreed in the Covenant.

Regarding the theory on self-regulation, the working group, which is a case of self-regulation with a regulatory framework imposed by an authority, reveals several issues. First, it should be acknowledged that self-regulation entails high initial transaction costs that none of the parties is ready to pay spontaneously. This is because it is a high risk in the initial phases of negotiation. It implies that policy making must either enforce the negotiations or pay for it. Secondly, the negotiations enable the exchange of knowledge and experience despite opposing interests and mistrust because the parties are forced to deal with each other as agreed in the Covenant. The result of such an exchange is dissemination of available solutions because the negotiators balance out. That is, they optimize, via deal making, the positive and negative effects of the solutions that advance their own interests. The negotiations occur as if they are a marketplace for the demonstration of new solutions, which underpins the theory. The possibilities to foster, and even encourage, innovations are limited unless new coalitions are formed that introduce new products due to regulatory pressure from the agreement. That is, the threat of strict regulations in the case of non-compliance, which is not in line with the theoretical assumptions. So far, it has been the first step in the negotiations process.

Following the working group negotiations, the Board and the ministries realized that many innovative options had not been used because of barriers such as legal limits for use of recycled materials to pack foods, too much focus on returnable packaging and others. The possibilities of other innovative, approaches have been assessed by a study and in a workshop with roughly 30 experts from industries, retail and environmental organizations, with policy makers in the role of observer. The study did not reveal new options in addition to the previous assessments and working group analyses. The results of the workshop, however, have revealed the different perceptions and interests of the suppliers (producers and packers) and the demanders (retail and consumers) regarding innovative options in packaging. The differences obstruct the introduction of innovations.

Workshop participants were divided into two groups: (a) packaging suppliers (producers, packers and researchers), (b) the demanders of packaging (retail, consumers and environmental organizations). The inventory rounds in two separate groups generated many ideas for innovative packaging. Following that, both groups agreed on 12 key innovative clusters for packing. The key clusters are:

1. New packing machines, such as flexible and module-based machinery;
2. Better design of packaging, such as balloon, vacuum-packing by consumers and so on;
3. Better quality of products, such as less perishable food, less fragile electronic products;
4. Normalization of the form and volume of packed products, such as equal size and portions;
5. Standardization of packaging materials that is mono-material;
6. Optimizing durability that is a better balance between market demand and product quality;
7. Telecommunication use in distribution, such as electronic selection of products in retail;
8. Reduction of packaging functions, e.g. separation of promotion and transport functions;
9. Production closer to consumers, e.g. meal preparation in shops, adding ingredients and diluting;
10. Adaptation of packaging materials towards less fragile, renewable resources;
11. Adaptation of presentation functions, such as new billboards and better use of shelves;
12. Optimizing the disposal and processing of packaging waste.

Thereafter, the groups evaluated separately the clusters with respect to attractiveness and market impediments for implementation. Four criteria for attractiveness have been agreed: (a) environmental relevance, (b) contribution to economy, (c) contribution to changes in consumer behavior, (d) contribution to changes in producer behavior. Four agreed criteria for the impediments are: (a) political-social impediments, (b) techno-organizational impediments, (c) resistance of vested interests, (d) consumer resistance. Every group evaluated each criterion using the values 0, 1 or 2. The total score per cluster is noted in the quadrants found in Table 9.2. The quadrants are divided into scores of 1–6 points and 7–12 points. Table 9.2 shows the results of the evaluation: vertically the attractiveness is rated, horizontally the impediments. The cluster numbers are presented in the scheme. The priority clusters for the co-operative strategy should be in the quadrant with the highly attractive clusters and high impediments, because joint efforts to develop the innovations are needed. Individual companies can focus on the quadrant of highly attractive clusters with low impediments.

Table 9.2. Innovative clusters in packaging assessed by suppliers and demanders of packaging

(Agreement is given in bold)	Low impediments, 1–6 points		High impediments, 7–12 points	
	Suppliers	Demanders	Suppliers	Demanders
Unattractive, 1–6 points	7		5 9	4 7
Attractive, 7–12 points	1 8 11	3 12	2 3 4 6 7 10	1 2 5 6 8 9 10 11

Only the cluster "telecommunication" (cluster 7) was found to be unattractive by suppliers and demanders. All of the other clusters were found attractive by demanders or suppliers, but high impediments were found as well (quadrant 4). The groups agreed on attractiveness in 9 out of 12 clusters. The non-attractive ones included: the normalization of products (cluster 4) was found attractive by suppliers but not by the demanders, the standardization of materials (cluster 5) was found attractive by demanders but not by suppliers. Moving production closer to consumers (cluster 9) was found unattractive by suppliers and attractive by demanders. An even larger difference was found in the perception of impediments. Only three clusters were prioritized for co-operative actions: packaging design (cluster 2), optimizing durability (cluster 6) and new packing machines (cluster 10). The analysis of the opinions revealed that if the quadrant would be drawn at 7 to 12 points, which is in fact in strict accordance with the methodology, then no one priority cluster could be found for co-operative actions. The method and the results have been evaluated by the participants after the meeting and generally supported. The workshop and the discussions thereafter revealed that no consensus could be reached on any of the innovative options. Neither interventions by the Board, nor funding for innovations from the ministries was enough to push the co-operation any further. Innovations did emerge, but through bilateral co-operation between individual suppliers and customers, and without the involvement of other stakeholders in the chain.

The negotiations and workshop on packaging illustrate that market negotiations about tangible demands provide benefits in comparison with a command and control policy. However, the results worsen in cases of ambiguous demands. Two main benefits can be identified: (a) it is possible to speed up diffusion of available technologies entailing product adaptations to comply with environmental demands, (b) it is possible to achieve targeted emissions reduction at costs close to the use of market-based instruments because the interests negotiate costs and benefits amongst them. Stakeholder negotiations as a market place for the available environmental technologies are feasible under the conditions of rigid targets and clear liability for non-compliance. This is imposed upon the private interests by an authority. The negotiations about environmental innovations failed, except in one case in which an informal coalition comprising environmental groups and a supplier worked together under the threat of regulation for non-compliance. A potential benefit based on the innovations was not achieved. Despite much research, the innovations discussed in stakeholder negotiations were found to be too laborious regarding differences in perceptions among stakeholders in the chain.

9.4 Model for Self-regulation

The case of the Packaging Covenant can be generalized. Three interests can be distinguished: a demander of environmental improvements, a polluter and an innovator. Hence, a simple negotiation model is presented to assess conditions for environmental innovations. In a model, a group of demanders expresses its interests through the readiness to impose liability on the polluter, if necessary, in court. When confronted with the demand, the emission sources can implement available technologies at increasing costs as a function of the demand for emissions reduction. The perspective of increasing costs triggers suppliers of environmental technologies to develop and deliver cost-reducing innovations. However, beforehand they must invest in the innovations. Realistic are the assumption that the demanders' interests are not uniform and instead depend on individual socio-economic positions, that these differences are known to the polluters and used to secure their interests and that the innovators are uncertain (before investing in technology development) whether the polluters are willing and able to buy the innovations. We abstract from competition between the polluters and between the innovators. A key element in the negotiation, we argue, is uncertainty about the demand for and possibility to sell the innovations.

The uncertainties involved in the negotiations can be illustrated by a realistic situation at an airport (based on work for Fokker, a airplane company that was triggered by customers' demands to limit airplanes noise profile in 1980s, with successful product development but sales failure due to airport waivers for incompliance with the noise regulation; Fokker went bancrupt in 1990s). The focus is on the noise made by airplanes. Three key interests can be distinguished: the citizens who live in the surrounding area (demanders), the airline companies or the airport authority as their representative (emission sources) and the airplane manufacturers (technology suppliers). The citizens press for less noise that is damaging to their health and welfare. However, the demands are ambiguous because the distance between the noisy airport and people's housing varies and many local residents work at the airport and want to keep their jobs. It is uncertain what is demanded, whether the citizens want to press for legal action or want to pay for less noise. The airlines rightly expect that the pressure to reduce noise decreases when fewer citizens are affected by noise. It implies uncertain demands as a decreasing function of the emissions-reduction percentage. The airlines can reduce noise by having noisy planes make fewer flights, change the position of airstrips and offer financial compensation, which means using available technology. Such a strategy strongly increases the costs of the existing fleet as a function of the progressing demands. The alternative is to negotiate with manufacturers of airplanes about a much lower noise profile that we call heroically "silent planes", which means using an innovation. The consequence of the latter is that manufacturers have to invest in the development of "silent planes" beforehand, that is, without the assurance that airlines will buy the new planes. The airlines, however, are not willing to assure the procurement because available technologies can be cheaper in the short run, even though they may be more costly due to the stricter demands of the citizens. The suppliers' investment in technology development therefore depends on airlines' interest in buying the innovation. This, in turn, depends on citizens' demands.

The benefits of the innovations are that (a) citizens gain due to noise reduction, (b) airlines reduce costs through "silent airplanes", (c) airplane manufacturers gain profit from the sales. The social benefit is expressed by the innovation rent, which is the present value of the users' expected savings plus the innovators' expected profit. In this case, it is the airlines' savings plus the manufacturers' profits. The incentive to innovate increases as the innovation rent grows. Distribution of the innovation rent is relevant but it can be neglected as long as the innovators expect sufficient profits. For convenience, we assume that the total innovation rent is accrued by the suppliers. The larger the innovation rent, the more likely the success of the negotiations, but success remains uncertain. The model is a simplification. Firstly, the citizen's monetary benefits are not covered. For example, real-estate prices may rise due to the reduced level of noise. Secondly, competition between the users and the suppliers of innovations as well as distribution of innovation rent are neglected. Above all, innovation spin-off is beyond our scope, although the effects of such an innovation on social awareness, corporate responsibility and the diffusion of new technologies are important arguments in favor of the innovations. For example, the spin-off of "silent airplanes" could be that citizens living near other airports urge the use of "silent airplanes" because they are already available in other places. This, in turn, benefits the citizens and provides a competitive advantage to the innovator. In summary, the positive effects of such negotiation results are underscored in the model.

Two types of uncertainty involved in such a negotiation are distinguished. The first type of uncertainty is whether the demanders enforce or compensate for noise reduction and to what extent because the issue becomes less pressing as the nuisance declines. Conversely, additional payments are needed to compensate for additional noise reduction. If the polluters cannot reduce the uncertainty about the demands they may pay excessively for emissions reduction or underscore the demand. This type of uncertainty can be reduced by further negotiations but at increasing transaction costs and delay in the procurement of the innovation. However, this reduces the present value of their savings, the innovation rent, similarly to the situation in policy making. Such negotiations are modeled to assess whether self-regulation provides a stronger incentives for innovations than policy regulations (Dosi and Moretto, 1997). On the assumption that emission sources innovate, it is argued that self-regulation requires lower costs than policy making and that a welfare optimum can be achieved. However, self-regulation is less certain than policy regulation that can be enforced. Therefore, emission sources discount profit at a higher interest in a situation of self-regulation than in policy regulations. Providing a subsidy for innovating polluters is recommended to compensate for the extra uncertainty present compared to policy regulation that is equal to the extra costs of a higher interest. The second type of uncertainty, which is missed by the study above, relates to negotiations on sales taking place between emission sources that use innovations and the technology developer. The developers are uncertain whether the emission sources will buy the innovations or use available technologies. If the suppliers overestimate sales, they will produce at a loss. If there is no guarantee about sales, then the profit of investment in technology development is uncertain (note that a guarantee does not end uncertainty about dissemination).

Table 9.3 shows the main variables and factors that influence the innovation rent. The uncertainties are approximated by use of a higher interest for discounting savings and sales revenue. The differences between self-regulation and policy making include more uncertain demand for innovations and uncertain sales, but there is no waiting period and enforcement is fast.

Table 9.3. Variables and factors that influence innovation rent

Cost saving by emission source	The demand for emissions reduction
	The production costs of available technologies
	Uncertainty about demand (interest)
	Investment in adaptation of the available technologies
Profit from innovation sales by suppliers	Sales revenue
	The production costs of innovations
	Uncertain sales (interest)
	Investment in R&DD

The effects of self-regulation on innovation rents are simulated and compared to the effects of policy making with an extra 4 years waiting time, t_w. The data are presented in Chapter 8. To summarize them: the expected sales, v, are 150 units per period (total in simulation 1 500 units), the production costs, o, are 50% of the sales, the investment, I, is 100 units, the enforcement period, t_e, is 15 years.

Three situations are compared: policy making, self-regulation with use of the available technologies and self-regulation with use of the innovations. In the case of self-regulation, it is assumed that the emission sources have a choice to invest in adaptations of available technologies or in innovations, thus saving money. They discount the expected costs savings due to adaptations or innovations with regard to uncertain demands. The self-regulation without waiting time is assumed to be more risky than policy making, which is reflected in a 5% higher interest rate. It is in line with the Dosi and Moretto model (1997). The additional perspective is one of the innovators. The innovators that can be technology suppliers as well as an R&D department in a polluting industry, invest in innovations based on expectations about the sales to emission sources (possible clients) but face the possibility of emission sources' adaptations. It is assumed that the innovators discount the expected cost savings at emission sources at 10% interest rate.

The results of simulations are compared with the reference that is 0% interest rate in policy making. In addition, the surplus of innovation rents in the case of the policy making is compared with the surplus in the case of the self-regulation with adapations and with innovations to indicate the (dis)advantages of self-regulation. The basic data are shown in Appendix E. Table 9.4 shows the results of the simulation. The simulation is:

In policy making: $t = t_e + 4$ for $r_{p_i} = 100\%, 105\%, 110\%, 115\%, 120\%$.

$$S_p(r) = \{\frac{\sum_{t=1}^{n}[V_{(t)} - O_{(t)}]}{r_{p(i)}^{t}}\} - I \qquad (9.1)$$

$$Index(p) = \frac{S_p(r_{p_i})}{S_p(r_{p_1})} \text{ (column 2)} \qquad (9.2)$$

For self-regulation and adaptations $t = t_e + 0$ and $r_a = r_p + 5\%$

$$S_a(r) = \{\frac{\sum_{i=1}^{n}[V_{(t)} - O_{(t)}]}{r_{s(i)}^{t}}\} - I \qquad (9.3)$$

$$Index(a) = \frac{S_a(r_{s_i})}{S_a(r_{p_1})} \text{ (column 3)} \qquad (9.4)$$

For self-regulation and innovations $r_i = 110\%$

$$S_i(r) = \frac{\sum_{i=1}^{n} S_{a_i}}{r_{s(i)}^{t}} - I \qquad (9.5)$$

$$Index(s) = \frac{S_i(r_{i_j})}{S_i(r_{i_1})} \ \text{(column 4)} \tag{9.6}$$

Benefit of self-regulation with innovations

$$B_s = \frac{S_{i(r)}}{S_{p(r)}} \ \text{(column 5)} \tag{9.7}$$

Table 9.4. Comparison of the surplus of innovation rent in policy making with waiting time and self-regulation with a higher interest rate.

	Policy making with 4 years waiting time	Self-regulation with interest rate 5% higher than in policy making; technology suppliers discount the savings at 10% interest rate		
1	2	3	4	5
Interest policy making	4 years extra waiting period	Emission sources' present value of savings	Technology suppliers' innovation rent	Benefit of self-regulation to policy making
100%	0.43	0.28	0.12	0.29
105%	0.22	0.19	0.08	0.38
110%	0.11	0.13	0.06	0.52
115%	0.04	0.09	0.04	0.80
120%	0.008	0.06	0.02	2.53

The policy making with a moderate waiting time of four years compared to self-regulation outweighs the surplus of emission sources' adaptations and even more the innovators' surplus because the self-regulation is risky. A long waiting time, however, counteracts this advantage. This holds true for the low interest in policy making, which means that policy making can ensure use of the innovation, for example using economic instruments. As the use of innovations through policy making becomes more uncertain or the waiting period involved in policy is extensive, self-regulation becomes more attractive for the innovators, both, for emission sources and for technology suppliers. The simulation shows, firstly, a trade-off between the waiting period in policy making and the higher uncertainty involved in self-regulation. The shorter the waiting period, the larger surplus in policy making compared to self-regulation because of lower uncertainty. Secondly, it shows the importance of innovations for the risky policy making that is in cases of expected strong resistance to any policy making, inability to prepare environmental demands, expected difficulties with enforcement and others. In such highly controversial situations, self-regulation within a regulatory framework is thus a "second best" option for innovations.

The risks for innovators, however, are also high. Regulations by authority are needed. There are several types of instruments that can enlarge innovation rents under self-regulation, and thereby foster innovations. One type of policy instrument is to strengthen the position of demanding environmental groups. This can be done through extended liability for products and non-compliance, emission rights and amenity rights, compensation for damages, take-back regulations and other forms of liability. Another type can reduce the uncertainty surrounding procurement of innovations. These instruments can be volume, price and quality guarantees such as quantum procurement by authorities, labels and certificates. The third type of policy instrument is a subsidy that can be divided into three categories: subsidies for technology development (for innovators), subsidies to reduce the transaction costs of negotiations (for emission sources or for environmental groups) and subsidies for procurement of innovations (note that the subsidy for procurement of available technologies reduces the innovation rents, thus it discourages innovations). The simulations indicate that policy instruments that reduce uncertainty about demand provide the strongest incentives to invest in innovations. The advantages of self-regulation in comparison with policy making largely depend on demanders' possibilities to make polluters liable. Subsidies reduce the costs but not the negative effects of uncertainty about the innovation rent. Therefore, subsidies are less effective than instruments to reduce uncertainty. The most effective one is a subsidy for technology development, particularly for the controversial and risky issues.

9.5 Conclusion

Self-regulation within a regulatory framework is an alternative to policy making. However, self-regulation is highly disputed. Its proponents underline positive attitudes from industries and possibilities to adjust to demands for emissions reduction at low costs. Opponents argue that self-regulation is not feasible because of the common good character of environmental qualities, diversity of interests in product chains, negotiations' transaction costs and more. An attempt to introduce self-regulation within the regulatory framework of an agreement between policy makers and industries on prevention and recycling of packaging waste reveals that product adaptations made through negotiations are feasible, but that innovations are hard to achieve because of the diverging interests. Two main benefits are identified: faster diffusion of available technologies with product adaptations and the achievement of targeted emissions reduction at the costs close to market conformity albeit at a high transaction cost and delay. This occurs under the conditions of clear targets and liability for non-compliance imposed upon private interests by an authority. Negotiations on environmental innovations failed because of differences in perceptions and the interests of stakeholders involved in the chain. Innovations' potential social benefits are not achieved.

In light of these experiences, conditions are discussed for successful stakeholder negotiations on innovations taking into consideration the demanders of emissions reduction, emission sources and innovators. It considers the possibilities of demanders to enforce the demands and benefit from emissions reduction, as well as innovators' profitable sales from new technology. Innovations can be pursued within a regulatory framework with liabilities despite uncertainties about the enforcement of the demands and the sales of innovations compared to use of available technologies. The choice between policy making and self-regulation within a regulatory framework is essentially a deliberation between less-certain self-regulation and a longer waiting period in policy making before innovations can be sold. Simulations of the innovation rents as a result of policy making and self-regulation suggest that the high uncertainty of self-regulation reduces innovation rents but it is not worse than the long waiting periods involved in policy making. Self-regulation is more attractive in risky areas within environmental policy. The effective policy instruments that support self-regulation in such areas include strong assurance of the demands by a regulatory framework of liability and compensation, as well as by price and quality guarantees.

10

How to Progress?

The general question is whether it is possible to reduce emissions substantially at progressively lower social costs. The currently available environmental technologies already make it possible to reduce most emissions by 70%, sometimes even as much as 95%. More effective technology is not the main prerequisite for enduring socio-economic progress and far-reaching emissions reduction. The challenge is to develop and disseminate effective technologies at continuously decreasing costs. For this purpose, environmental innovations are needed.

The empirical findings about resource use show a trend towards reduced use of resources per unit of output and steadily decreasing resource prices. This trend is mainly due to increased services. Similar to trends in resource use, one can advocate technological development towards a higher emissions-reduction percentage at lower unit costs. Theories on technology development suggest that steering towards low-cost pollution reduction is possible but different steering mechanisms are proposed. Neoclassical theory argues that higher prices on pollution are needed to reduce emissions. There is no doubt that price is a strong incentive for emissions reduction. Yet how can one guarantee a steady price increase to reduce emissions progressively when past trends indicate that prices are not the only factor involved in resource-saving technology development, and possibly not even the most crucial one? Evolutionary theory suggests that we are locked in technological patterns that cause pollution, so policy makers need to break out of these old patterns. Sound policy making is obviously needed, but how can it avoid misconceptions in decision making on environmental qualities and technological development? Behavioral theory pinpoints the role played by vested interests that inhibit innovation processes. Forceful, external demands are needed to focus managerial decisions on environmental issues, like the oil shock occurring in the 1970s that triggered energy saving, or consumer actions in the 1980s that sparked ecodesign. In this view, changes of interest drive innovations. However, how can these changes be fostered? A key factor is to reduce uncertainty in decisions about environmental innovations. Those demanding emissions reduction are uncertain about its social benefits and how to impose environmental demands. Polluters are uncertain what demands should be expected and whether new technologies are better than available ones. Innovators are uncertain how profitable their innovations will be at the moment of decision making about a costly investment in technology development.

10.1 Innovations Reduce Costs

Apparently, policy makers can make firm decisions to enforce strict environmental demands as needed from an environmental quality perspective. However, in reality, these decisions may not distort economic structures or income growth. So policy makers tend to avoid the risk of placing demands that are beyond the scope of available environmental technologies from the past in a business branch. This is because they expect that it can lead to unreasonably high costs. Policies prepare demands for emissions reduction based on the best available technologies that are assumed to cost not much more than the average cost in that business branch. The problem arises as to how to assess that stricter demands will not cause a large cost burden, because at the time of policy preparation there is hardly any experience with the best available technologies as technology progresses. Practical experience is lacking because companies are not interested in investing in environmental technologies spontaneously. During policy preparation, knowledge about the costs and effects of advancing technologies can only be generated by executing a few demonstration projects at emission sources. For a reliable demonstration project, a method is needed to indicate which source is representative for many other types and sizes of emission sources in terms of costs of emissions reduction.

Engineers often argue that the costs of a technology in an inventory with similar emission sources like a business branch, is determined by the scale of emissions reduction. Tests with empirical data on technologies for emission sources that are provided by engineers and consultants show that the relationship between costs and the scale of emissions reduction is weak. The costs are not related to the scale for even one type of technology at several emission sources in one company, or for widely used pollution-control equipment in a branch. The cost escalation is not robust enough to construct reliable cost functions of emissions reduction. Demonstration projects at a large emission source and at a small one, even at a few more sources of different scales in a branch, are insufficiently reliable to assess the costs at various other emission sources in that branch. This is because process variables largely determine the costs of emissions reduction. The alternative for the escalation is to take for granted that the cost functions of emissions reduction cannot be defined in detail. The "second best" option is to indicate the cost line without detailed information about every individual source. The cost lines of the empirical cost functions of emissions reduction can be described by the streamlined cost functions that relate the unit costs to each other. It is formally: $c_{i+1} = c_i * e^k$. The c_{i+1} and c_i are unit costs of consecutive emission sources on the streamlined cost function. The coefficient k is the exponent of the natural logarithm, which indicates how steep the cost function is: the larger the cost exponent, the steeper the cost functions, so the greater the increase of unit costs. The way to construct an accurate streamlined cost function is by using emission sources with very low unit costs and very high unit costs, not a large and a small source. Thereafter, the unit costs and cumulative percentage of emissions reduction can be linked. It is found that by doing this way the streamlined cost functions are accurately constructed for the majority of 28 empirical cost functions of emissions reduction ($R^2 > 0.9$ for 21 out of 28). All other linear and non-linear methods to approximate empirical cost functions provide inaccurate outcomes. The limitation of the streamlined cost functions is that the unit costs and the scale of emission reduction at the sources cannot be assessed. This is because there is no direct link between the scale and the costs. The accuracy of a streamlined cost function can be explained by the non-parametric statistical theory on the probability of events, which is described by Poisson distribution. The streamlined cost function represents the probability of finding an environmental technology that matches with emissions reduction at a subsequent source in the inventory. Gradually increasing unit costs implies a high probability of finding a suitable technology at sources in the inventory. Strongly increasing unit costs means low probability, which indicates that it is difficult to find the match. The theory on industrial loss prevention underpins the argument that two factors are likely to determine the unit costs: the large number of variables in the process means costly handling and immaturity of the technology implies few cost-saving adaptations. A complex and immature process cause high unit costs at a source. The implication of the analysis for policy making is that it is not possible to assess the costs of emissions reduction at every source. It is only possible to define a streamlined cost function. The latter is useful to indicate the costs and effects of economic instruments, such as a charge or a subsidy, because the price on emissions can be related to the emissions-reduction percentage.

Even less is known about the costs and benefits of environmental innovations when preparing environmental policy. During preparation, the innovator must assess whether investment in research, development, demonstration and manufacturing of new technologies can be covered by profitable sales of the innovation. Sales at a profit can be expected if polluting companies that are users of the innovation save money in comparison to using available technologies. In view of possible savings, the choice is to invest in realization of innovations or adapt the available technologies. It is found that innovations are attractive in the inventories of emission sources and technologies that can be described by the streamlined cost functions with a large cost exponent and adaptations at the flat streamlined cost functions. For 28 streamlined cost functions, the turning point of innovation versus adaptation is found at a cost exponent (k_c) above 0.24 with about 20% spread. For inventories with the steep streamlined cost functions (k_c above 0.24), it is attractive to develop a new technology. The strongly increasing costs of emissions reduction suggest that niches for innovations can be found. In the areas of acidification and heavy-metal emissions into air, it is rather unattractive to innovate because the innovations must compete with many available technologies and the sales are likely to be low in comparison with investments in technology development. In other areas, such as greenhouse gases and volatile organic compounds released into air as well as phosphates and heavy metals released into water, innovations are likely to be more profitable. In this way, the streamlined cost functions can be used to prioritize funding of environmental innovations in the early phases of policy preparation. Based on the estimates about the attractiveness of innovation it is assessed whether higher investments in innovations can be justified by a high social benefit. The assessment is based on empirical data in the Netherlands. The social benefit of the cost-saving innovations is due to profitability of the sales of the new technologies by innovators and because the polluting companies can save costs during the life cycle of the installed technologies. The assessment shows that strict policy provides a sufficiently large market for the profitable sales to enlarge technology development. The key areas for the sales of innovations are energy saving and renewable energy. Most important for the social benefit are the potential cost savings in the polluting industries that can be many times higher than the expenditures in technology development, even under conservative assumptions. The policy demands that do not invoke innovations can unintentionally cause unnecessary high costs in the polluting industries because possible cost savings during the life cycle of equipment are underused. Good assessments of the life-cycle costs, therefore, encourage profitable innovations.

Turning to polluting companies, it is discussed whether the costs of compliance with policy demands impede productivity growth as economists usually assume. This economic viewpoint has been tested with statistical materials from the period 1980–2002, using data for refineries, chemical industries, basic metal and the electrical power sector on reduction of the acidifying air emissions and for food sector and refineries with chemical industries on the biodegradable emissions to water. The analysis shows that companies can reduce acidifying emissions released into air by 60% to 84% alongside decreasing unit costs by 18% to as much as 23% per year. This means that unit costs are cut in half compared to the initial ones in 4 and 3 years, respectively. Similar, though less impressive, technological progress is found in the food and refineries with chemical industries with respect to emissions reduction of biodegradable matter. This was reduced by 70% to 80% at decreasing costs of emissions reduction of 8% to 14% a year. The findings indicate that industries are able to cut down emissions and costs. Technological progress is attained through investments and even more through improved use of technologies. The latter underlines the importance of capable environmental management and workers in the polluting industries. The findings shows that there are ample opportunities for cost savings through environmental innovations entailing adaptations during use. The high rate of effect-increasing and cost-reducing technological progress implies that the effects of strict environmental policy on productivity become negligible after a few years of enforcement. In addition, there is reason to expect some benefits from environmental innovations due to material and energy saving and better products as a side effect of innovative, environmental management.

The explanation of the findings about the cost reduction in the polluting industries is based on a decision model for environmental management. In the model environmental management has the choice of selecting the best available technology as defined by policy makers and adopting innovations that can find a solution suitable to the specific company's situation. The innovation requires company expenditures to search, select and test the technology before implementation that are called the cost of change-over. This is a burden in the short-term, but such a cost of using innovation is possibly more rewarding in the long term under strict environmental demands than the use of the available technologies because the latter would need costly upgrading to comply with the stricter demands. In addition, an innovation can yield energy and material savings as well as sales of better products. Such a dilemma between the cost of change-over for innovation in the short term and long-term benefit is encountered in many investment decisions. The issue in decision making of environmental managers is uncertainty about whether such strict policy demands are going to be implemented. That is, is it worthwhile to cover the costs of change-over and adopt environmental innovations or is it better to wait and implement available technologies? Case studies illustrate that the choice for innovations can contribute to positive company results in some areas of environmental management albeit the positive side effects of using innovations are usually outbalanced by the costs of change-over and expenditures in other areas of environmental management.

In addition to policy demands, companies increasingly encounter social demands from stakeholders including consumer and environmental organizations that pursue environment-oriented products with consideration of products' life cycle. One difficulty in decision making is that stakeholders' demands often contradict product functionalities demanded by customers. As a result, companies must operate as if there were two markets with different prices and qualities. Another difficulty is the complexity of life cycles. Many methods have been developed to support decision making but all are imperfect compared to a theoretical full pricing of pollution. An approximation of the pricing of pollution by life-cycle costing reveals that many innovative opportunities can be found to reduce the additional costs. It suggests that life-cycle management can reduce the cost burden imposed by stakeholders' demands. Cases of life-cycle management show that the potential, additional costs of far-reaching emissions reduction to attain low impacts on the environment can be prevented by a few focused actions in the chain. One or two focused actions neutralize 60% to 80% of the potential, additional costs of emission reduction and in some cases, generate a benefit due to better product marketing and resource savings.

10.2 Conditions for Cost-saving Innovations

Analyses reveal that companies' performance can substantially be improved through emvironmental innovations. The question, therefore, is how to foster progress on cost-reducing environmental technologies? The issue is to define the conditions that enable innovators to develop and sell profitably environmental innovations. In the instrument theory, it is widely assumed that environmental authorities induce innovation through strict demands, if necessary, by forcing polluting companies to implement technologies still unavailable at the moment of policy making. In reality, however, policy makers seek credible enforcement at reasonable costs. Technology-forcing regulations are rare and examples do not provide convincing results but rather delay of implementation by companies. A more realistic assumption is that policy makers prepare and enforce demands based on credible information about demonstrated technologies. Innovators that demonstrate new low-cost technologies ultimately determine progress in policy demands. The viewpoint of innovators is therefore relevant for far-reaching emissions reduction.

According to the instrument theory, a distinction is usually made between market-based policy with economic instruments, such as charges and direct policy with emission standards in permits. Covenants are seen as the instruments in the middle. Economists generally assume that given a demand, economic instruments provide the strongest incentive for innovations because companies pay for residual emissions. Based on the innovation theory that addresses the profitability of innovations, it should be argued that the preferable policy instruments reduce the costs of innovation, increase the volume of possible sales and reduce the uncertainty surrounding the preparation and enforcement of the demands. The choice of policy instrument is relevant insofar as it contributes to these factors. The economic instruments can reduce the costs of innovations through charges and subsidies. These expand sales volume as reduction of residual emissions pays and reduce uncertainty because only the polluting companies decide which technology is going to be implemented instead of holding negotiations with authorities in the case of permits.

The impediment for profitable innovations is the lead time and uncertainty involved in policy making. As soon as an environmental problem is signaled and receives attention, innovators can start to develop technologies. During policy preparation, innovators must demonstrate that their technology can be used to comply with environmental demands. Then they must wait many years until the demands are set by politicians and enforcement starts. This is because polluting companies have no interest in implementing such new technology spontaneously, except for those companies that take the risk of anticipating enforcement because they can save costs or improve their products. The waiting period facing the innovator between the moment of demonstration and enforcement is long, usually lasting between six to eight years. The waiting time of covenants is shorter. It is longer in the case of economic instruments due to high resistance. The start of enforcement enables the sale of technologies. The growth of sales on environmental technology rarely expands rapidly as the sales are bound to negotiations between authorities and companies. There is also limited administrative capacity and uncertainty about whether innovative, low-cost solutions are going to be implemented or rather the solutions proposed by vested interests. Innovation suppliers are confronted with uncertain enforcement that usually takes more than a decade. Economic instruments provide an advantage compared with other instruments. They offer freedom to choose technology and an incentive to act quickly because it reduces payments. The short waiting period and low uncertainties during preparation and enforcement are key factors for the profitable environmental innovations. Laborious and uncertain policy making causes the net present value of future sales to decrease. For example, an eight-year waiting period, followed by eight years of enforcement at an interest reate of 10% causes only 5% of the sales value to remain compared to an immediate implementation of a successful innovation, which enables the sales of technologies. Better conditions for environmental innovations can be created by the timely announcement of environmental demands, followed by fast preparation and enforcement. Economic instruments are effective if prepared in a short period of time. Subsidy for emission reduction is another option.

An alternative approach is self-regulation carried out through negotiations between stakeholders. In this approach, policy makers create favorable conditions for the negotiations, such as distributing the rights to claim good environmental qualities or emissions rights. In theory, it is expected that the negotiations achieve optimal solutions under the conditions of timeless and costless negotiations. The assumptions of the theorem have been criticized for a lack of realism, but the negotiations are realistic with additional policy instruments such as liability or tradeable rights. In comparison with policy making, negotiations offer the advantages of no waiting period, as is found in policy preparation, and that enforcement depends solely on market interests. The disadvantage is that negotiations cause more uncertainty about enforcement of demands. Seeking representation of the interest groups cannot overcome every difficulty, because it remains uncertain if the representatives are able to impose the demands or to pay for emissions reduction. An example of negotiations in packaging chains illustrates that there is a great potential benefit of innovative solutions in the chain, but it is difficult to reach an agreement on priorities for environmental innovations, let alone to start implementation. The barriers are different perceptions and interests of suppliers of packaging and demanders of packed products. Under self-regulation, two types of uncertainty related to the sales of innovations emerge. The management of polluting companies is uncertain about stakeholders' demands and innovators are uncertain if the polluters will buy available technologies or the innovations. The possibilities for innovations through negotiations must take into account the potential savings of the polluting companies and profits for the innovator, with consideration of uncertainties about sales. There is a trade-off between the waiting period in policy preparation and the uncertainties involved in negotiations. Beforehand, there is no best choice. However, the negotiations are risky for innovators unless policy instruments are introduced as well. The most effective instruments reduce uncertainty about environmental demands by introducing liability for damage, emission rights and so on. It is also possible to reduce uncertainty about sales through procurement, price and quality guarantees for innovations, producers' responsibilities in the chain and so on. Subsidies are less effective than these instruments to reduce uncertainty. An effective subsidy is for technology development.

The question of whether it is possible to reduce emissions at socially acceptable costs has been positively answered. There is technology to attain a far-reaching emissions-reduction percentage if that is demanded by policy makers and market interests. The costs of available technologies decrease due to adaptations based on experiences in use. Environmental innovations provide substantial cost reductions and they can even contribute to corporate results. Innovations can be triggered by environmental demands in policy making with a short lead time in policy preparation and rapid enforcement. They can also be promoted through negotiations between stakeholders if harmed interests obtain legal instruments to assure that their demands can be enforced. Policies that embark on environmental innovations and adaptations reduce emissions down to levels that do not preclude sustainable environmental qualities at a cost that hardly increases, and may even bring about cost savings in many cases.

A

Appendix to Chapter 3

Table A.1. Review of the environmental technologies in the inventories, n is the number source technology combinations

Sector-wise inventories
Benzene (n = 13) technologies: 1. Adapted seals, 2. Floating roofs, 3. Isolation (PUR, aluminum), 4. Thermal combustion, 5. Stripping and condensing, 6. Maintenance and control, 7. Vapor return and condensation, 8. Coal filter.
Cadmium (n = 4), technologies: 1. Cloth filter, 2. Process change phosphoric acid, 3. Filters after combustion.
CO_2 (n = 79), technologies: 1. Co-generation, 2. High-efficiency boilers, 3. Membrane chloride production (diaphragm substitution) 4.Optimizer building heating, 5. Efficient agriculture machines, 6. Improvement prereformer, 7. Membrane technology, 8. Process integration with heat pumps, 9. Reuse energy fly-wheel trains, 10. Washing machines hot fill, 11. Derivates, 12. Isolations, 13. Heat tanks, 14. Furnace carts, 15. Wind energy, 16. Solar energy for drying, 17. SiC: preheating with CH_4, 18. Dry coke quenching, 19. HDPE: Fluidized bed process, 20. Adapted steel ovens at rolling, 21. Protein winning, 22. Procurement CO_2, 23. Earth heat, 24. Olefins: selective steam cracking, 25. Combustion and gasification wood, 26. Water power, 27. Hydrogen winning with membrane, 28. Diesel high effectiveness, 29. Biogas manure, 30. Saving lamps, 31. Low-energy greenhouse.
Phenol (n = 7), technologies: 1. Faster substitution catalysts, 2. Better maintenance, 3. Gas washing, 4. Catalytic combustion, 5. Optimal water use, 6. Thermic combustion.
Fine dust (n = 13), technologies: 1. Cloth filters, 2. Electrostatic filters, 3. Closed silos 4. Filters for diesel exhaust.
Fluoride (n = 8), technologies: 1. Recycling glass, 2. Dry treatment with CaO, 3. Adsorption at aluin earth processing, 4. Gas washing, 5. Clean phosphoric acid process, 6. Electrofilter.

Table A.1. (continued)

Sector-wise inventories, continued
Phosphate (n = 10), technologies: 1. Biological water treatment, 2. Dephosphating, 3. Manure processing.
Copper (n = 4), technologies: 1. Closed docks, 2. Clean phosphoric acid process, 3. Sludge combustion and filtering, 4. Clean pigment process.
Metals to air (n = 13), technologies: 1. Unleaded petrol, 2. Filtering exhaust
Metals to water (n = 16), technologies: 1. Biological water treatment, 2. Low-copper fodder, 3. Recycling of heavy metals
NH_3 (n = 12), technologies: 1. Low-NH_3 stables, 2. Isolation manure storage, 3. Two-phase fodder, 4. Filtering (venturi or bio filter), 5.Direct plough-in manure.
NOx (n = 102), technologies: 1. Low NO_x burner, 2. non–selective catalytic reduction, 3. Process adaptation ammonia production, 4. Process adaptation nitric acid production, 5. Process adaptation caprolactam production, 6. Water injection, 7. Engine tuning, 8. Selective catalytic reduction, 9. Autocatalyst, 10. Premix in carburettor, 11. Premix at combustion
PAH (n = 6), technologies: 1. Low temperature wood conservation, 2. Adsorption in water treatment, 3. Optimization of cokes production, 4. LPG use, 5. Conversion of multi-burners to coal burners.
Propylene oxide (n = 7), technologies: 1. Gaswashing, 2. Maintenance and control, 3. Vapor cooling, 4. New seals, 5. Rings on pumps and isolation of tanks.
SO_2 (n = 36), technologies: 1. Adaptation soot emission, 2. Tailgas unit, 3. CaO injection, 4. Process emission building materials, 5. Process emission aluminium, 6. Substitution coal by gas, 7. Oil and diesel desulfurization.
Styrene (n = 15), technologies: 1. Biofiltration, 2. Active coals and regeneration, 3. Maintenance and control, 4. Low-styrene resins, 5. Better seals, 6. Catalytic combustion, 7. Thermical combustion.
Toluene (n = 20), technologies: 1. Absorption and reuse, 2. Thermical combustion, 3. Catalytic combustion, 4. Maintenance and control, 5. Isolation tanks, 6. Low-toluene primer, 7. Floating roofs on tanks, 8. Closure valves in coke chamber, 9. Balance injection in tanks.

Table A.1. (continued)

Sector-wise inventories, continued
VOC (n = 47), technologies: 1. Vapor return and floating roofs, 2. Cooling machines, 3. VOC-low print of PVC, 4. VOC-low washing and care, 5. VOC-free undercoat, 6. Combustion and condensation, 7. VOC-low paint, 8. Pesticide reduction, 9. VOC-free ink, 10. Balance injection in pipes, 11. Condensation and absorption TRI, 12. Vapor return in gasoline tanks and pumps, 13. Maintenance and control, 14. Reversed osmosis, 15. Biofiltration
Zinc (n= 6), technologies: 1. Recycling (membrane), 2. Neutralization, 3. Biological treatment, 4. Cleaner zinc production, 5. Cleaner chlorine process
Company-specific inventories
Chlorinated-organic (n = 23), technologies: 1. Vapor degreasers, 2. Water and soap, 3. Screw wax, 4. Spray tunnel, 5. Soak bath, 6. Three step degreasing, 7. Ultrasonic soak, 8. Ceramic polishing, 9. Closed rotating bath.
NO_x chemical industry (n = 256), technologies: 1. Low NO_x burners, 2. Gas turbine, 3. Furnace, 4. Absorption, 5. Chimney adaptation, 6. Selective catalytic reduction, 7. Process-adaptation nitrate, 8. Process-adaptation nitric acid, 9. Air preheating, 10. Boiler adapted fireplace.
NO_x electric power (n = 100), technologies: 1. Process optimization, 2. HTNR-burner, 3. Low NO_x burner, 4. Ring combustion chamber, 5. Adapted fireplace, 6. Various combustion, 7. Selective catalytic reduction.
NO_x basic metal (n = 73), technologies: 1. Gas burners, 2. Adapted boiler, 3. Adapted heating 4. Fire stores, 5. Fluor washing, 6. Chimney adaptation, 7. Furnace adaptation, 8. Process adaptation ovens, 9. Co-generation, 10. Electrolyse process optimization.
NO_x refineries (n = 41), technologies: 1. Low-NO_x burner, 2. Electrostatic precipitation with selective catalytic reduction, 3. Selective catalytic reduction
SO_2 chemical industry (n = 56), technologies: 1. Gas washing, 2. Hydroxide washing, 3. Hydrogenation, 4. Scrubber, 5. Degussa, 6. Gas for diesel
SO_2 electric power (n = 19), technologies: 1. Fluegas desulfurisation, 2. Gas desulfurisation at fireplace.
SO_2 basic metal (n = 44), technologies: 1. Gas washing, 2. Hydroxide washing, 3. Water spray, 4. Injection CaO, 5. Absorption.
SO_2 refineries (n= 18), technologies: 1. Electrostatic precipitation/hydroxy washer, 2. Gas for oil

Constructions of the streamlined cost function, example of fluoride.
Three scenarios for fluoride emission reduction into air are elaborated. The data are known about only two source technology combinations:

1. Scale and unit costs of combinations are based on the highest and lowest unit costs: $k_c = \ln (c_8/e_1)/(n-1) = \ln (691 / 1.4) / 7 = 0.89$, so $c_2 = 1.4 * 2.7180^{89}$ etc.; $k_{e1} = \ln (e_8/e_1) / (n-1) = \ln (30\ 000 / 18\ 000) / 7 = 0.073$, so $e_2 = 30\ 000 * 2.7180^{0.073}$, etc.

2. Scale and unit costs of combinations are based on the largest and smallest scale: $k_2 = \ln (c_7/e_3)/ (n-1) = \ln (207 / 8.3) / 7 = 0.46$, so $c_2 = 8.3 * 2.71800^{0.46}$ etc.; with $k_e = \ln (e_3/e_7) / (n-1) = \ln (570\ 000 / 1\ 000) / 7 = 0.89$, so $e_2 = 1\ 000 * 2.7180^{89}$ etc.

3. The data are known about two source technology combinations, the scale and rank of the scale. Three variants are calculated:

    a. All data are known: $k1 = [\ln (c_{i+1}/c_i) + \ln (c_{i+2}/c_{i+1}) +...+\ln (c_n/c_{n-1})] / (n-1)$. For fluoride $n = 3$: $c_1 = 1.4$, $c_2 = 6.9$, $c_3 = 8.3$, $k1 = (\ln [(6.9 / 1.4) + \ln (8.3 / 6.9)] / 2 = 0.90$ so, $c_4 = c_3 * 2.7180^{90}$.

    b. Randomly selected two combinations: $k2 = \ln (c_{i+m}/c_i) / (n-1)$. In fluoride: $c_4 = 19.3$, $c_3 = 8.3$, $k2 = (19.3 / 8.3) / (4-3) = 0.86$, so, $c_5 = c_4 * 2.7180^{0.86}$ and $c_2 = c_3 2.7180^{-0.86}$.

    c. The highest cost and the lowest unit cost combination: $k3 = \ln (c_n/c_i) / (n-1)$. In fluoride: $c_8 = 691$, $c_i = 1.4$, $k3 = \ln (691 /1.4) / 7 = 0.89$ so, $c_2 = c_1 * 2.7180^{89}$.

Table A.2 shows the estimate for the third scenario (all data are known). The rank changes for the exponents that are based on the interpolation of the largest and smallest combination (column 2). Accuracy (R^2) and reliability (C) are accounted for and presented at the bottom of the table.

Table A.2. Fluoride example of streamlined cost function

Empirical and streamline fluoride data; costs in € and €/kg, n_{kc} = unit cost rank, n_{ke}= scale rank										
Ranks	empirical data						all data known	random nrs. 3 & 4		highest– lowest unit
n_{kc}	n_{ke}	e real	e hypoth	C		c	c_1 $k1=$ 0.9	C_1/C	c_2 $k2=$ 0.86	C_2/C	c_3 $k3=$ 0.89	C_3/C

n_{kc}	n_{ke}	e real	e hypoth	C	c	c_1	C_1/C	c_2	C_2/C	c_3	C_3/C
1	6	18 000	18 000	19 800	1.4	1,4	100	1.4	104	1.4	100
2	3	33 000	19 400	230 415	6.9	3,7	50	3.2	49	3.2	48
3	1	570 000	20 800	4 608 295	8.3	8,3	105	8.3	100	8.3	101
4	7	12 000	22 400	230 415	19.4	21	108	19.4	100	19.8	103
5	5	26 000	24 100	1 152 074	44	50	113	45.6	103	48.3	109
6	2	120 000	25 900	9 216 590	77	118	159	108.3	141	117	152
7	8	1 000	27 900	207 373	207	297	143	257	124	284	137
8	4	30 000	30 000	20 737 327	691	721	104	611	88	691	100
Total		810 000	189 000	3 6407 373		$R^2=$ 0.99	$C_1=$ 118	$R^2=$ 0.99	$C_2=$ 104	$R^2=$ 0.99	$C_3=$ 114

Table A.3. Exponents, accuracy and reliability of the streamlined cost functions (*str*) compared to the empirical cost function (*emp*) based on the combinations with the lowest and the highest unit costs

Inventories	n	cr_l	cr_n	kcr	$R^2 cr$	er_l	er_n	ker	$R^2 er$	Er emp/str	Cr emp/str
Sector-wide											
Benzene	13	0.86	987	0.59	0.97	8 000	6 900	−0.011	0.96	5.4	6.5
Cadmium	4	150	57 604	1.98	0.96	39	420	0.776	0.99	1.1	1.2
CO_2	79	.0003	1.4	0.10	0.85	123 088 694	12 125 000	−0.029	0.96	4.8	5.5
Copper	4	76	1 260	0.94	0.91	27 200	65 200	0.285	0.94	0.6	0.8
Fine dust	13	0.52	9	0.24	0.97	84 ,000	15 489 000	0.237	0.94	0.5	0.5
Fluoride	8	1.4	691	0.89	0.99	18 000	30 300	0.073	0.85	4.3	1.1
Phosphate	10	1.9	68	0.40	0.98	5 350 000	1 576 000	−0.133	0.93	1.6	1.0
Metals air	13	82	1 198	0.22	0.97	6 000	40 000	0.155	0.61	6.4	1.6
Metals water	14	11	614	0.27	0.94	570 000	1 600	−0.442	0.96	1.6	5.7
NH_3	10	1.7	14	0.19	0.88	5 275 910	5 945 000	0.013	0.99	2.0	1.3
NO_x	99	0.07	23	0.06	0.92	87 580 000	15 484 000	−0.017	0.96	0.1	0.2
PAH's	6	84	822	0.46	0.95	256 000	110 000	−0.165	0.83	0.5	0.5
Phenol	7	0.92	53	0.68	0.97	5 000	2 000	−0.153	0.90	5.5	8.4
Propylene	7	0.87	27	0.57	0.98	31 800	8600	−0.218	0.99	0.9	0.8
SO_2	35	0.22	3	0.08	0.91	4 972 500	60 000	−0.126	0.88	21.6	28.2
Styrene	15	0.98	35	0.26	0.97	567 000	433 000	−0.019	0.96	0.5	0.6
Toluene	20	0.03	461	0.51	0.96	150 000	15 000	−0.121	0.96	2.0	5.3
VOC	47	0.30	7	0.07	1.00	1 300 000	6 400 000	0.035	0.94	1.1	0.7
Zinc	6	46	3 383	0.86	0.98	17 100	3 800	−0.291	0.97	0.9	1.4
Company–specific											
Cl–metal	23	1.7	656	0.27	0.98	20 800	25	−0.306	0.98	1.5	2.2
NO_x chemic	253	0.02	1 005	0.04	0.90	5 050	780	−0.007	0.98	65.7	4.9
NO_x electric	97	0.14	99	0.07	0.83	301 084	20 725	−0.027	0.99	13.3	6.4
NO_x metal	70	0.10	16 109	0.17	0.57	5 488	3	−0.108	0.69	125.9	216.4
NO_x refinery	39	0.43	878	0.19	0.65	188 000	90	−0.191	0.98	12.3	33.5
SO_2 chemic.	53	0.59	2 166	0.15	0.93	1 495 200	150	−0.167	0.96	1.1	1.4
SO_2 electric	17	0.35	5	0.15	0.96	28 224 342	568 958	−0.217	0.99	1.6	1.2
SO_2 metal	43	1.2	188	0.12	0.82	2 614 305	9 000	−0.132	0.96	0.6	0.7
SO_2 refinery	17	1.2	31	0.19	0.92	4 312 800	39 600	−0.276	0.98	2.5	2.0

B

Appendix to Chapter 4

Simulated adaptations and innovations, example of fluoride.
For the fluoride inventory, the costs after adaptation are:
$c_1 = 1.4 * 0.1, c_2\,7 * 0.23... c_8 = 691 * 1$
the total savings after simulation are: $1 - (30\,285\,240 / 36\,407\,373 * 100\%) = 17\%$
the costs after innovation are: $c_1 = 1.4 * 1, c_2 = 7 * 0.87...c_8 = 691 * 0.1$
the total savings after simulations are: $1 - (9\,762\,870 / 36\,407\,373) * 100\%. = 73\%$

For fluoride emission reduction, it is worthwhile to innovate because the savings due to the innovations are larger than the savings due to the adaptations. In a similar way, all other streamlined cost functions are elaborated.

Table B.1. Simulations of adaptations and innovations for fluoride inventory (costs in €)

n	empirical data			adaptation policy			innovation policy		
	E	C	C	% empiric	c	C	% empiric	c	C
1	18 000	1.4	24 885	10%	0.14	2 488	100%	1.4	24 885
2	33 000	7	230 415	23%	1.6	52 666	87%	6.1	200 790
3	570 000	8.1	4 608 295	36%	2.9	1 645 820	74%	6.0	3 423 305
4	12 000	19	230 415	49%	9.3	111 916	61%	12	141 540
5	26 000	44	1 152 074	61%	27	707 702	49%	22	559 579
6	120 000	77	9 216 590	74%	57	6 846 610	36%	27	32 91 639
7	1 000	207	207 373	87%	181	180 711	23%	47	47 400
8	30 000	691	20 737 327	100%	691	20 737 327	10%	69	20 73 733
Total			36 407 373			30 285 240			97 62 870
Saving						83%			27%
k_c		0.89			1.22			0.56	

Table B.2. Numerical values of savings by adaptations and innovations, single emissions

Cost functions	n	k_c	Total costs: simulated/empiric		Innovation /Adaptation
			Adaptation	Innovation	
NO$_x$ chemical	256	0.04	0.68	0.42	0.61
NO$_x$	102	0.06	0.72	0.38	0.53
NO$_x$ electric power	100	0.07	0.69	0.41	0.60
VOC	47	0.07	0.74	0.36	0.49
SO$_2$	36	0.08	0.32	0.78	2.44
CO$_2$	82	0.10	0.75	0.35	0.46
SO$_2$ metal	44	0.12	0.52	0.58	1.10
SO$_2$ electric power	19	0.15	0.42	0.68	1.63
SO$_2$ chemical	56	0.15	0.54	0.56	1.04
NO$_x$ metal	73	0.17	0.75	0.34	0.45
SO$_2$ refineries	18	0.19	0.42	0.68	1.65
NO$_x$ refineries	41	0.19	0.38	0.72	1.88
NH$_3$	12	0.19	0.65	0.45	0.68
Metals air	13	0.22	0.55	0.55	1.00
Fine dust	13	0.24	0.93	0.17	0.18
Styrene	15	0.26	0.81	0.29	0.36
Metals water	16	0.27	0.50	0.46	0.91
Cl-hydrocarbons	23	0.27	0.50	0.60	1.20
Phosphate	10	0.40	0.71	0.39	0.55
PAH	6	0.46	0.81	0.29	0.36
Toluene	20	0.51	0.86	0.24	0.28
Propylene oxide	7	0.57	0.72	0.38	0.54
Benzene	13	0.59	0.90	0.17	0.19
Phenol	7	0.68	0.69	0.41	0.59
Zinc	6	0.86	0.84	0.26	0.31
Fluoride	8	0.89	0.83	0.27	0.32
Copper	4	0.94	0.91	0.19	0.21
Cadmium	4	1.98	0.93	0.17	0.18

Table B.3. Numerical values of savings by adaptations and innovations, clusters

Cost functions	n	k_c	Total costs: simulated/empiric		Innovation /Adaptation
Correlations			**0.55**	**−0,54**	**−0.44**
< 0.1	5	0.06	0.63	0.47	0.94
0.1–0.2	8	0.16	0.55	0.54	1.11
0.2–0.4	5	0.25	0.66	0.41	0.73
0.4–0.6	5	0.50	0.80	0.29	0.38
0.6–0.9	3	0.81	0.79	0.31	0.41
> 0.9	2	1.46	0.92	0.18	0.20
Correlations			**0.92**	**−0.91**	**−0.87**

Abbreviation of emissions names used in Tables B.4 and B.5

Be	Benzene
Ca	Cadmium
CO_2	Carbondioxide
Phe	Phenols
Flu	Fluorides
Pho	Phosphates
Cop	Copper
Mea	Heavy Metals to air
Mew	Heavy metals to water
NH_3	Ammonia
PAH	Polycyclic aromatic hydrocarbons
Pro	Propylene Oxide
SO_2	Sulfur dioxide
Dus	Fine dust
Sty	Styrene
Tol	Toluene
Vo	Volatile organic compounds
Zin	Zinc

Table B.4. Sales possibilities of environmental innovations based on attractiveness of innovations (D) for the investments in environmental technology (€ mln)

E/D	Be 83%	Ca 83%	CO₂ 65%	Ph 59%	Flu 73%	Pho 61%	Cop 81%	Mea 45%	Mew 54%	NH₃ 55%	NOx 62%	PAH 71%	Pro 62%	SO₂ 22%	Dus 83%	Sty 71%	Tol 76%	VO 64%	Zin 74%
1	0.2	0	17	0.1	0.0	11	7	1	7	16	15	10	0.1	0.2	1	1	0.2	1	4
2	0.4	0	27	0.1	1	12	8	1	8	18	18	11	0.1	0.3	1	1	0.3	1	5
3	1	0	45	0.1	1	13	9	1	8	20	23	13	0.1	0.5	2	2	0.4	2	6
4	1	0	74	0.2	1	13	10	2	9	22	28	15	0.1	1	3	2	1	2	6
5	2	0	121	0.2	1	14	12	2	9	25	35	18	0.1	1	4	2	1	3	8
6	2	0	199	0.2	2	15	13	3	9	27	44	21	0.1	1	6	3	2	4	9
7	4	0	327	0.2	2	16	15	3	10	31	54	24	0.1	2	9	3	3	5	10
8	6	1	537	0.2	3	17	17	4	10	34	68	28	0.1	3	13	4	5	7	12
9	10	1	882	0.2	4	17	20	5	11	38	84	32	0.1	4	18	4	9	10	14
10	17	2	1 450	0.3	6	18	23	7	11	42	105	37	0.1	5	27	5	14	14	16
11	27	3	2 383	0.3	8	19	26	9	12	46	131	43	0.1	7	40	5	24	19	19
12	43	5	3 915	0.3	10	20	30	11	13	51	163	50	0.1	10	59	6	39	27	22
13	70	8	6 433	0.3	14	22	34	13	13	57	203	58	0.1	14	86	7	65	37	26
14	112	13	10 571	0.4	19	23	39	17	14	63	253	67	0.1	20	127	8	106	51	30
15	180	22	17 371	0.4	26	24	45	21	14	70	315	77	0.1	29	186	10	176	70	35

Table B.5. Cost savings 2% a year during the life cycle of technology use. Depreciation on average 21 years based on civil 60% in 25 years, mechanical 40% in 15 years (€ million)

	Be	Ca	CO2	Ph	Flu	Pho	Cop	Mea	Mew	NH3	NOx	PAH	Pro	SO2	Dus	Sty	Tol	VO	Zin
1	0	0	529	0	1	218	44	10	131	194	124	90	1	23	9	12	0	8	16
2	0	0	700	0	1	247	50	15	145	225	167	101	1	33	13	14	0	12	19
3	1	0	926	0	1	279	57	21	160	262	225	113	1	48	18	17	0	18	22
4	1	1	1226	0	2	316	65	30	177	306	304	126	1	68	25	20	1	27	26
5	2	1	1622	0	3	359	74	42	196	356	411	142	1	97	35	24	1	41	30
6	3	2	2146	0	4	406	84	60	217	414	555	159	1	139	49	29	3	60	35
7	6	3	2839	0	7	460	96	86	240	483	749	178	1	199	69	35	5	90	41
8	12	6	3756	1	10	521	109	122	265	562	1011	199	1	285	97	41	11	134	47
9	23	10	4970	1	15	590	124	173	293	654	1365	223	1	407	135	50	21	200	55
10	43	17	6576	1	23	669	142	247	325	762	1843	250	1	582	190	60	41	298	64
11	80	30	8700	1	35	757	161	351	359	887	2488	280	1	833	266	71	80	444	74
12	151	52	11511	1	54	858	184	499	397	1033	3360	314	1	1191	372	86	158	662	86
13	284	90	15230	2	82	972	210	710	440	1203	4536	351	1	1703	522	103	309	987	100
14	534	155	20151	2	124	1101	239	1010	487	1401	6125	393	1	2437	731	123	605	1471	116
15	1004	268	26662	2	405	1247	359	1436	538	1632	8269	441	2	3485	1025	148	1185	2192	135

C

Appendix to Chapter 7

Example of life-cycle costing: RME and diesel

This example of the RME account illustrates the accounting procedure:
Table C.1 Compliance costs to produce tonne of RME
Table C.2 Compliance costs to produce tonne of Diesel
Table C.3 Compliance costs in use of Diesel and RME
Table C.4 Production and compliance costs of Diesel and RME

Table C.1. Compliance costs per tonne RME, excl. byproducts of straw and glycerol
Production costs about € 1000 per tonne

Steps	Emission kg	Reduction %	Reduction kg	Unit costs € /kg	Tot costs €/tonne
Rape growing	2 500				
N	10	83	5	2.5	10.5
P	13	80	10	5.5	55
Pesticides	5.62	90	5	5	26
N₂O P.M.					
CH₄ P.M.					
Pesticides	0.56				
NOₓ	0.83	38	0	9.5	3
SO₂	0.35	55	0	1.7	0,5
CO₂	50	54	27	0,5	14
				Total pollution control costs	109

Table C.1. (continued)

Steps	Emission kg	Reduction %	Reduction kg	Unit costs €/kg	Tot costs €/tonne
Rape seed oil	1 000				
Fodder	1 500				
Hexane	3.9	83	3	4	12.5
CO_2	164	50	82	0.45	39
SO_2	0.693	72	0	0.7	0.5
NO_x	0.334	87	0	7	2
NH_3 P.M.					
Total pollution control costs					154
Esterification	1 000				
Total pollution control costs to produce 1 tonne RME					163

Table C.2. Pollution control costs in production per ton of diesel

Emissions	Emission kg	Reduction %	Reduction kg	Unit costs €/kg	Total costs in €
To air					
CH_4	2.2				
CO_2	14.5	66	9.6	0.05	0.3
NO_x	0.1	88	0.1	2.5	0.2
SO_2	18.0	84	15.1	0.7	11
VOC	6.9	69	4.8	4.9	23
Costs emission reduction to air					34
To water					
COD	0.0	100	0.0	0.4	0.0
Cl-hydrocarbons	0.0	83	0.0	129	0.0
Heavy metals	0.0	96	0.0	428	0.4
PAHs	0.0	99	0.0	573	0.0
Organic phenol	0.8	39	0.3	67	17
Costs emission reduction to water					17.4
Waste					
Chemical	0.4	100	0.4	1.3	0.
Industrial	17.8	100	17.8	0.4	3.2
Costs of waste treatment					3.5
Total pollution control costs					55.5

Table C.3. Pollution control costs in use of diesel and RME for 253 000 km

User phase	Emission (kg/km)	Reduction (%)	Unit costs €/kg	Total costs in €
Diesel (13.6 tonne in the life cycle)				
VOC	0.00031	75	6	324
NO$_x$	0.00084	50	12	2 245
CO	0.00140	0		
Particles	0,00038	65		
SO$_2$	0.00024	95	4.3	249
CO$_2$ (*)	0.20100	49	0.3	7 911
Pb	0.00000	0		0
VOC from tank	0.00006	95	9	125
Pollution control costs in use				10 875
RME (14.1 tonne in the life cycle)				
VOC	0.00028	0		
NO$_x$	0.00109	50	21	2 918
CO	0.00126	0		
Particles	0.00018	0		
SO$_2$	0.00000	0		
CO$_2$	0.19497	49	0.3	7 674
Pb	0.00000	0		
VOC from tank	0.00006	0		
Pollution control costs in use				10 593
(*) based on fuel saving by technical adaptations of cars				

Table C.4. Life-cycle costs of diesel and RME for 253 000 km

		Production cost	Use costs	Total excl. tax	Taxes	Total, incl. taxes
Diesel	Fuel	1 635		1 635	8 920	10 554
	Emission	75.5	10 875	11 631		11 631
	Total	2 390	10 875	13 265	8 920	22 185
RME	Fuel	6 965		6 964	1 529	8 494
	Emission	2 306	10 593	12 899		12 899
	Total	18 540	10 593	19 863	1 529	21 392
RME advantage		−6 880	283	−6 597		
Loss of taxes					7 390	

Abbreviation of emissions names used in Tables C.5 and C.6

Be	Benzene
Ca	Cadmium
ClO	Chlorinated organic compounds
CO_2	Carbon dioxide
Phe	Phenols
Flu	Fluorides
Pho	Phosphates
Cop	Copper
Mea	Heavy metals to air
Mew	Heavy metals to water
Nit	Nitrates
NH_3	Ammonia
PAH	Polycyclic aromatic hydrocarbons
Pest	Pesticides
Pro	Propylene Oxide
SO_2	Sulfur dioxide
Dus	Fine dust
Sty	Styrene
Tol	Toluene
Vo	Volatile organic compounds
Zin	Zinc
Hwa	Household waste
Owa	Organic waste
Waw	Hazardous waste
Sha	Small hazardous waste (e.g. batteries)
Iwa	Inorganic waste (e.g. demolition waste)

Abbreviation of emissions sources names

Agri	Agriculture
Food	Food industries
Text	Textile, leather and garments industries
Print	Printing and paper industries
Ref	Refineries and petrochemical industries
Chem	Chemical industries
Met	Basic metal industries
Equ	Metal products and machine industries
Electr	Electrical and electronic industries
Cons	Construction and building
Serv	Services
Dist	Distribution and storage
Load	Cargo transport
Per	Personal transport
Hous	Households
Waste	Waste processing
Other	
Gener	General

Table C.5. Strict environmental demands (D) per sector and in general; % pollution reduction compared with 1990

	Agri	Food	Text	Print	Ref	Chem	Metal	Equip	Electr	Const.	Serve	Dist	Load	Pers	House	Waste	Other	Gener
Ben					57%	19%	27%					85%	35%	35%	90%			66%
Ca						98%	53%		82%									66%
ClO	90%														83%			83%
CO$_2$	54%	49%	50%	52%	66%	50%	62%	60%	70%	50%	40%		50%	50%	60%		50%	59%
COD																		100%
Ph										39%								39%
Dus						99%	70%	51%				77%	77%			90%		72%
Flu						5%	26%			90%								32%
Pho	80%	90%				90%		90%							65%			97%
Cop						83%		40%					55%			100%		35%
Mea				80%		80%	80%	80%				100%			100%	100%		98%
Mew	53%			54%		93%	70%	51%	75%		75%					96%		72%
NH$_3$	49%					44%				50%								62%
Nit															58%			83%
NO$_x$	38%	73%	85%	89%	88%	87%	75%	88%	92%	52%	63%		51%	79%	77%		84%	77%

Table C.5. (continued)

R max	Agri	Food	Text	Print	Ref	Chem	Metal	Equip	Electr	Const.	Serve	Dist	Load	Pers	House	Waste	Other	Gener
Pest	90%																	50%
PAH							65%			80%				90%	99%			85%
Pro					73%													73%
SO$_2$	55%	71%	70%	70%	84%	72%	73%	70%	95%	78%	68%		41%	41%	60%		70%	87%
Sty						94%								90%				79%
VO	70%	31%	70%	64%	69%	83%		60%		54%		69%		90%	41%			73%
Tol			58%	37%		20%	21%	99%		42%		52%		92%				41%
Zinc			95%			40%		84%										80%
Hwa															100%		100%	100%
Owa															100%		100%	100%
Haw																	100%	100%
Sha															100%			100%
Iwa									100%							100%		100%

Table C.6. Long term maximum unit costs of pollution control to comply with the strict environmental demands per sector and in general €/kg

	Agri	Food	Text	Print	Ref	Chem	Met	Equip	Elect	Cons	Serve	Dist	Load	Pers.	Hous	Waste	Other	Gener
Ben					10	8	26					688	16		538			688
Cad						162	25 000		40 116									40 116
ClO	7							55							129			129
CO₂	0.52	0.94	0.04	0.05	0.03	0.43	0.04	0.08	0.14	0.02	0.22		0.07	0.01	0.75		−0.02	0.95
COD																		
Phe										58								58
Dus						2	6		3				2	161		1		161
Flu						225	481			48								481
Pho	5	4				6	6	9							48			48
Cop								643					578			877		877
Mea				693		702	312	539				89			89	834		834
Mew	67			359		12	180	345	279		428					351		428
NH₃	10					7				4								10
Nit															1			2
NO$_x$	9.5	9	7.5		2.5	7	7.7	9.5	4	7	7		3.5	5.5	16.5		14	16.5

Table C.6. (continued)

cr_n	Agri	Food	Text	Print	Ref	Chem	Met	Equip	Elect	Cons	Serve	Dist	Load	Pers.	Hous	Waste	Other	Gener
Pest	5																	5
PAH							250			91				326	572			572
Pro					16													16
SO₂	1.6	0.8	0.85	0.75	0.7	0.7	0.75	0.9	0.5	1.7	1.33	1.33		2.3	2.3		0.9	2.3
Sty						38								592				592
VO	0.5	5	1.5	1	5	9		1.5		3.5		5		1.5	5			5
Tol			0.5	5		137	50	8		55		341		209				521
Zin			50			2 356		1 390										2 356
Hwa															0.22		0.12	0.22
Owa															0.25		0.02	0.25
Haw																	0.65	0.65
Sha															2.5		2.5	2.5
Iwa									0.01							0.18	0.18	0.18

D

Appendix to Chapter 8

Table D.1 is divided into three parts, each one shows one type of enforcement. One finds the undiscounted sales and the discounted innovation rent for the waiting time of 0 years, 4 years and 8 years and 15 years of enforcement, no R&D.

Table D.1. Simulation of sales and innovation rents of policy enforcement

Year	Gradual enforcement						Slow enforcement						Fast enforcement					
	Sales			Rent			Sales			Rent			Sales			Rent		
	0	4	8	0	4	8	0	4	8	0	4	8	0	4	8	0	4	8
1	100			73			2			1			500			227		
2	100			41			3			1			334			138		
3	100			38			4			1			223			84		
4	100			34			6			2			149			51		
5	100	100		31	31		9	2		3	1		99	500		31	155	
6	100	100		28	28		13	3		4	1		66	334		19	94	
7	100	100		26	26		20	4		5	1		44	223		11	57	
8	100	100		23	23		29	6		7	1		29	149		7	35	
9	100	100	100	21	21	21	44	9	2	9	2	0	20	99	500	4	21	106
10	100	100	100	19	19	19	66	13	3	13	3	1	13	66	334	3	13	64
11	100	100	100	18	18	18	99	20	4	17	3	1	9	44	223	2	8	39
12	100	100	100	16	16	16	149	29	6	24	5	1	6	29	149	1	5	24
13	100	100	100	14	14	14	223	44	9	32	6	1	4	20	99	1	3	14
14	100	100	100	13	13	13	334	66	13	44	9	2	3	13	66	0	2	9
15	100	100	100	12	12	12	500	99	20	60	12	2	2	9	44	0	1	5
16		100	100		11	11		149	29		16	3		6	29		1	3
17		100	100		10	10		223	44		22	4		4	20		0	2
18		100	100		9	9		334	66		30	6		3	13		0	1
19		100	100		8	8		500	99		41	8		2	9		0	1
20			100			7			149			11			6			0
21			100			7			223			15			4			0
22			100			6			334			20			3			0
23			100			6			500			28			2			0
Tot	1 500	1 500	1 500	408	260	177	1500	1500	1500	223	152	104	1500	1500	1500	578	394	269

E

Appendix to Chapter 9

Comparison of policy making with 4 years waiting time and self-regulation under the uncertainties of the emission source about the demands of stakeholders and the uncertainties of the innovators about the decisions of the emission sources (15 years of implementation, 100 units per year).

Table E.1. Simulation of innovation rents in self-regulation

Period (years)	Sales per year and total	Policy making	Self-regulation	
		Rent innovation	Rent adaptations	Rent innovation
Discount		20%	25%	10% demands
1			45	41
2			41	34
3			38	28
4			34	23
5	100	20	31	19
6	100	17	28	16
7	100	14	26	13
8	100	12	23	11
9	100	10	21	9
10	100	8	19	7
11	100	7	18	6
12	100	6	16	5
13	100	5	14	4
14	100	4	13	3

Table E.1. (continued)

Period (years)	Sales per year and total	Policy making	Self-regulation	
		Rent innovation	Rent adaptations	Rent innovation
Discount		20%	25%	10%
15	100	3	12	3
16	100	3		
17	100	2		
18	100	2		
19	100	2		
total	1 500	113	380	224

References

[1]   Allen, P.M., (1988), Evolution, innovation and economics, in G. Dosi, C. Freeman, R. Nelson, G. Silverberg, L. Soete (eds.), *Technical Change and Economic Theory*, Pinter Publishers, London/New York, pp. 95–119.

[2]   Andreasen, N.C., (2005), *The Creative Brain*, Penguin Group, New York, USA.

[3]   Angel, P.D., (2000), Environmental Innovation and Regulation, in G.L. Clark, M.P. Feldman, M. Gertler (eds.), *The Oxford Handbook of Economic Geography*, Oxford University Press Inc., New York, pp. 607–623.

[4]   Apeldoorn, M.E. van, C.A. van der Heijden, F.X.R. van Leeuwen, (1986), *Criteriadocument styreen*, RIVM, rapportnr. 738513003.

[5]   Arentsen M.J., P.S. Hofman, (1996), *Technologie: Schone motor van economie*, Universiteit Twente, CSTM, pp. 26–33.

[6]   Arentsen M.J., V. Dinica, E. Marquart, (2001), *Innovating Innovation Policy. Rethinking Green Innovation Policy in Evolutionary Perspective*, Économies et Sociétés, Série Dynamique technologique et organisation, nr 4 : 563–583.

[7]   Arora, S., T.N. Cason, (1996), Why do firms volunteer to Exceed Environmental Regulations? Understanding Participation in EPA's 33/50 Program, *Land Economics*, Vol. 72, No. 4:413–432.

[8]   Arrow, K.J., (1962), The economic implications of learning by doing, *Review of Economic Studies*, 29: 153–177.

[9]   Arthur, W.B., Yu.M. Ermoliev, Yu.M. Kaniovski, (1987), Path–dependent processes and the emergence of macro–structure, *European Journal of Operational Research*, No. 30:294–303.

[10]   Arthur, W.B., (1989), Competing Technologies, Increasing Returns and Lock–In by Historical Events, *The Economic Journal*, Vol. 99: 116–131.

[11]   Arthur, W.B., (1990), Positive Feedback in the Economy, *Scientific American*, February: 80–85.

[12]   Ashford, N.A., G.R. Heaton, (1979), The effects of Health and Environmental Regulation on Technological Change in the Chemical Industry: Theory and Evidence, in T. Hill (ed.), *Federal Regulation and Chemical Innovation, American Chemical Society*, Washington DC, pp. 45–66.

[13]   Ashford, N.A., C. Ayers, R.F. Stone, (1985) Using Regulation to change the market for innovation, *The Harvard Environmental Law Review*, Vol. 9, No. 49: 419–466.

[14]   Ashford, N., (1996), The influence of information–based initiatives and negotiated Environmental Agreements on Technological Changes, in C. Carraro, F. Lévêque, *Voluntary Approaches in Environmental Policy*, Kluwer Academic Publishers, Dordrecht, pp. 137–150.

[15] Ashford, N.A., (2005), Government and Environmental Innovation in Europe and North America, K.M. Weber and J. Hemmelskamp (eds.), *Towards Environmental Innovation Systems*. Springer Press, Heidelberg, pp 159–174.

[16] Ayres, R.U., (1978), *Resources, Environment, and Economics*, Applications of the Material/Energy Balance Principle, A Wiley–Interscience Publication, New York.

[17] Ayres R.U., (1989), Industrial Metabolism, in J.H. Ausubel, H.E. Sladovich (eds.), *Technology and Environment*, National Academy Press, Washington DC, pp. 23–49.

[18] Ayres, R.U., (1997), The Kuznets curve and the life cycle analogy, *Structural Change and Economic Dynamics*, No. 8: 413–426.

[19] Balmann, A., M. Odening, H–P. Weikard, W. Brandes, (1996), Path–dependence without increasing returns to scale and network externalities, *Journal of Economic Behaviour and Organisation*, Vol. 29: 159–172.

[20] Barbera, A.J., V.D. McConnell, (1986), Effects of pollution control on industry productivity: A factor demand approach, *Journal of Industrial Economics*, Volume XXXV, No. 2: 161–172.

[21] Barde, J–P, (1999), *Economic Instruments for Pollution Control and Natural Resources, Management in OECD countries: A Survey*, Organisation for Economic Co–operation and Development, Paris, mimeo.

[22] Baumol, W.J., W.F. Oates, (1975), *The Theory of Environmental Policy*, Prentice–Hall, Englewood Cliffs, pp. 14–55.

[23] Becker, N., M.G. Baron, M. Shechter, (1993), Economic Instruments for Emission Abatement under Appreciable Technological Indivisibilities, *Environmental and Resource Economics*, No. 3: 263–284.

[24] Beek, van Th.A., (1997), *Financiële beoordeling van productinnovaties*, Kluwer Bedrijfsinformatie, Deventer.

[25] Benhaïm J. and P. Schembri, (1996), Technical Change: An Essential Variable in the Choice of a Sustainable Development Trajectory, S. Faucheaux, D. Pearce, J. Proops (eds.), *Models of Sustainable Development*, Edward Elgar Publishers, Cheltenham, pp.105–122.

[26] Benn, S., D. Dunphy, A. Griffits, (2007), Integrating human and ecological factors: a systematic approach to corporate sustainability, in D. Marinova, D. Annandale, J. Phillipmore, *The International Handbook on Environmental Technology Management*, Edward Elgar Publishers, Cheltenham, pp.222–240.

[27] Beije, P., (1998), *Technological change in the Modern Economy*, Edward Elgar Publishers, Cheltenham, UK.

[28] Berkel, van C.W.M., (1996), *Cleaner Production in Practice*, IVAM Environmental Research, Universiteit van Amsterdam, mimeo.

[29] Beus, de J., (1991), The Ecological Social Contract, in D.J. Kraan, R.J. in 't Veld (eds.), *Environmental Protection: Public or Private Choice*, Kluwer Academic Publishers, Dordrecht, pp. 181–205.

[30] Blok, K., E. Worrell, R.A.W. Albers, R.F.A. Cuelenaere, (1990), *Data on energy conservation techniques for the Netherlands* (en het ICARUS 2 model op diskette), Vakgroep Wetenschap en Samenleving, Universiteit Utrecht, mimeo.

[31] Bohm P., C.S. Russel, (1985), Comparative analysis of alternative policy instruments, in A.V. Kneese, J.L. Sweeney (eds.), *Handbook of Natural Resources and Energy Economics*, Vol. 1, Elsevier, North–Holland, pp. 375–460.

[32] Bressers, J.Th.A., (1983), *Beleidseffectiviteit en waterkwaliteitsbeleid – een bestuurskundig onderzoek*, TH Twente, Enschede, mimeo.

[33] Bressers, J.Th.A., (1995), Motieven voor zelfregulering, in J.Th.A. Bressers, T.J.N.M. de Bruijn, S.M.M. Kuks, K.R.D. Lulofs, *Milieumanagement, een systematische aanpak voor bedrijven en andere organisaties*, Milieubeleid, Samson HD Tjeenk Willink, pp. 13–25.

[34] Brezet, J., (1994), *Van prototype tot standaard; de diffusie van energiebesparende technologie*, Denhatex BV, Rotterdam, mimeo.

[35] Bridgen, P., (2003), *ISO 14000: Global and Company Perspective*, Environment International LTD, mimeo.

[36] Bringezu, S., (1997), Comparison of the Material Basis of Industrial Economies, in S. Bringezu, M. Fischer–Kowalski, R. Kleijn, V. Palm (eds.), *Analysis for Action: Support for Policy towards Sustainability by Material Flow Accounting, proceedings of the ConAccount Conference*, Wuppertal (Germany), pp. 57–66.

[37] Bruijn, S.M. de, R.J. Heintz, (1999), The environmental Kuznets curve hypothesis, in J.C.J.M. van den Bergh (ed.), *Handbook of Environmental and Resource Economics*, Edward Elgar, Massachusetts. pp. 656–677

[38] BSO, *Annual reports*, 1993 en 1994, mimeo.

[39] Calef, D., R. Goble, (2005), *The Allure of Technology, How France and California Promoted Electric Vehicles to Reduce Urban Air Pollution*, Fondazione Eni Enrico Mattei, Milan, mimeo.

[40] Carrao C., D. Sinicalco, (1992), Environmental Innovation Policy and International Competition, *Environmental and Resource Economics*, Vol. 2: 193–200.

[41] Carraro, C., D. Siniscalco, (1994), Environmental policy reconsidered: The role of technological innovation, *European Economic Review*, No. 39: 545–554.

[42] Cavendish, W., D. Anderson, (1994), Efficiency and Substitution in Pollution Abatement, *Oxford Economic Papers*, No. 46: 774–799.

[43] CBS, Centraal Bureau voor de Statistiek, *Kwartaalberichten Milieu, Statline*, 1986–2005.

[44] CBS, Centraal Bureau voor de Statistiek, *Statistisch Zakboek*, 1981–2005.

[45] CBS, Centraal Bureau voor de Statistiek, *Milieukosten van bedrijven, Statline*, 1980–2005.

[46] CBS, Centraal Bureau voor de Statistiek, *Kosten en financiering van milieubeheer, Statline*, 1980–2005.

[47] CBS, Centraal Bureau voor de Statistiek, *Speurwerk en Ontwikkeling*, 1980–1993.

[48] Cerin, P. (2006), Bringing economic opportunity in line with environmental influence; A discussion on the Coase theorem and the Porter and van der Linden hypothesis, *Ecological Economics*, Vol.56, Issue 2: 209–225.

[49] Chiesa, V., E. Giglioli, R. Manzini, (1999), R&D Corporate Planning: Selecting the Core Technological Competence, *Technology Analysis & Strategic Management*, Vol. 11, No. 2: 255–279.

[50] Christensen, P., (1991), Driving Forces, Increasing Returns and Ecological Sustainability, in R. Constanza (ed.), *Ecological Economics*, Columbia University Press, New York.

[51] Christensen, C.M., (2000), The Innovator's Dilemma, *Harper Business*, Boston, pp. 4–68.

[52] Clarke, S.F., N.J. Roome, (1995), Managing for Environmentally Sensitive Technology: Networks for Collaboration and Learning, *Technology Analysis & Strategic Management*, Vol. 7, No. 2: 191–215.

[53] Cleven, R.F.M.J., J.A. Janus, J.A. Annema, W. Slooff, *Integrated Criteria Document Zinc*, Report No. 710401028, August 1993, mimeo.

[54] Coase, R., (1972), The problem of Social Cost, reprint in R. Dorfman en N.S. Dorfman (eds.), *Economics of the Environment*, selected readings, W.W. Norton & Company, New York, pp. 142–171.

[55] Cohan, D. Gess, (1993), *Managing life–cycle costs*, Decision Focus Incorporated, Mountain View USA, mimeo.

[56] Colinsk, J., (1996), Why Bounded Rationality?, *Journal of Economic Literature*, Vol. XXXIV, June: 669–700.

[57] Conrad, K., D. Wastl, (1995), The Impact of Environmental Regulation on Productivity in German Industry, *Empirical Economics*, No. 20: 615–633.

[58] Coombs, R., P. Saviotti, V. Walsh, (1987), *Economics and Technological Change*, MacMillan Education, London..

[59] Cooper R.G., E. J. Kleinschmidt, (1991), New products: what separates winners from losers?, in J. Henry, D. Walker (eds.), *Managing Innovation*, Sage Publishers, London, pp. 127–140.

[60] Coopers & Lijbrand, Dijker, Van Dien, CIVI Consultancy, *Technologische Oplossingsrichtingen voor Milieuproblemen*, December 1992, mimeo.

[61] Cramer, J., J. Schot, (1993), Environmental Comakership Among Firms as a Cornerstone in the Striving for Sustainable Development, in K. Fischer, J. Schot (eds.), *Environmental Strategies for Industry*, Island Press, Washington DC, pp. 311–328.

[62] Cramer, J., (1997), *Milieumanagement: van 'fit' naar 'strech*, Jan van Arkel, Utrecht.

[63] Cyert, R.M., J.G. March, (1988), A Behavioural Theory of Organisational Objectives, in R.M. Cyert (ed.), *The Economic Theory of Organisation and the Firm*, New York University Press, New York, pp. 125–150.

[64] Cyert, R.M., J.G. March, (1968), *A Behavioral Theory of the Firm*, Englewood Cliffs, N.J. Prentice Hall, New York.

[65] Dales, J.H. (1972), Land, Water and Ownership, reprint in R. Dorfman, N.S. Dorfman (eds.), *Economics of the Environment*, selected readings, W.W. Norton & Company, New York, pp. 229–245.

[66] Daly, H., J.B. Cobb jr, (1994), *For the common good*, Beacon Press, Boston.

[67] Dasgupta, P.S., G.M. Heal, (1979), Economic theory and Exhaustible Resources, *Cambridge University Press*, Cambridge, pp. 173–192.

[68] David, P.A., 1975, *Technical Choice Innovation and Economic growth*, Cambridge University Press, New York.

[69] DeCanio, S.J., W.E.Watkins, (1998), Investment in Energy Efficiency: Do The Characteristics of Firms Matter?, *The Review of Economics and Statistics*, Vol. LXXX, No. 1: 95 –107

[70] Derksen, H.P.J., J. Krozer, (1996), Economische aandachtspunten in de uitvoering van Convenant KWS 2000, in H.B. Diepenmaat et al., *Aandachtspunten in de uitvoering van Convenant KWS 2000*, TNO–MEP, mimeo.

[71] Desimone, L., F. Popoff, Eco–efficiency, (1997), *The Business Link to Sustainable Development*, MIT Press, Cambridge, Massachusetts.

[72] Dhillon, B.S., (1989), *Life Cycle Costing*, OPA, Amsterdam.

[73] Dieleman, H., S. de Hoo, (1993), Towards Tailor–made progress of pollution prevention and cleaner production, results and implications of the PRISMA project, in K. Fischer, J. Schot, *Environmental strategies for industry*, Island Press, Washington DC.

[74] D'Iribarne, P., What kinds of alternative ways of life are possible?, in L. Uusitalo (ed), *Consumer Behavior and Environmental Quality*, Gower Publishing, Hants, 1983, p. 27–37.

[75] Dhillon, B.S., *Life Cycle Costing*, OPA, Amsterdam, 1989.

[76] Doelman P., J. Krozer, J. Schot, (1994), *Een verkenning van de mogelijkheden voor preventie in de milieuvergunningen*, Tebodin Consulting Engineers, Den Haag, 1991, mimeo.

[77] Doelman, P., *Life Cycle Cost of a Television Set*, Institute for Applied Environmental Economics (TME), The Hague, mimeo.

[78] Doelman, P., J. Krozer, (1994), *Sociaal–economische gevolgen van het Convenant Verpakkingen*, Instituut voor Toegepaste Milieueconomie, Den Haag, mimeo.

[79] Doelman, P., M. Freriks, B. Kothuis, J. Krozer, A. Peer, (1994), *Ketenbeheer in scheepvaart*, Instituut voor Toegepaste Milieueconomie, mimeo.
[80] Doelman, P., J. Krozer, (1995), *Milieuverbeteringen van keukenkastjes*, Instituut voor Toegepaste Milieueconomie, Den Haag, mimeo.
[81] Doelman, P., J. Krozer, (1996) *Milieukosten van kopieerapparaten*, Instituut voor Toegepaste Milieueconomie, Den Haag, mimeo.
[82] Dosi, G., L. Orsenigo, (1988), Coordination and transformation: an overview of structure, behaviours and change in evolutionary environments, in G. Dosi, C. Freeman, R. Nelson, G. Silverberg, L. Soete, (eds.), *Technical Change and Economic Theory*, Pinter Publishers, London/New York, p. 13–37.
[83] Dosi, G., (1997), Opportunity Incentives and the Collective Patterns of Technological Change', *The Economic Journal*, No. 107: 1530–1547.
[84] Dosi G., M. Moretto, (1997), Pollution accumulation and firm incentives, *Environmental and Resource Economics*, Vol. 10:285–300.
[85] Downing P.B., L.J. White, (1986), Innovation in Pollution Control, *Journal of Environmental Economics and Management*, No. 13: 18–29.
[86] Driel, P. van, J. Krozer, (1987), Innovatie preventief milieubeheer en schonere technologie, *Tijdschrift voor Politieke Economie*, jrg. 10, nr. 4: 227–244.
[87] Driel, P. van, J. Krozer, (1987), *Op zoek naar goede daden*, Landelijk Milieu Overleg, Utrecht, mimeo.
[88] Driel, P. van, J. Krozer, (1987), Innovation and preventive environmental policy, in F.J. Dietz, W.J.M. Heijman (eds.), *Environmental Policy in a Market Economy*, selected papers from the Congress Environmental Policy in a Market Economy, Wageningen, Netherlands, 8–11 September, pp. 92–115.
[89] Drury, C, *Management accounting for business decisions*, Thomson Learning, London, 2001.
[90] Duchin, F., G., de Lange, (1994), *The Future of The Environment*, Oxford University Press, New York.
[91] Duchin, F.F., A.E. Steenge, (1999), Input-output analysis, technology and the environment, in J.C.J.M. van den Bergh (ed.), *Handbook of Environmental and Resource Economics*, Edward Elgar, London, pp. 1037–1059.
[92] Dutilh, C.E., (1995), Mechanismen die het produktbeleid bepalen, in J.J. Bouma, J.M.D. Kosten, H.R.J. Vollenbergh (red.), *Milieurendement in theorie en praktijk*, Samson HD Tjeenk Willink, Alphen aan de Rijn, pp. 153–159.
[93] Eads, G.C., (1990), Regulation and Technical Change: Some Largely Unexplored Influences, *Innovation and Technological Progress*, Vol. 70, No. 2: 50–54.
[94] Earl, G. (1999), Removing emotion from the environment – A multi–attribute stakeholder wide approach to resolving conflicts in company decision making in M. Backman and R. Thun (eds.), *Total cost assessment*, IIIEE Communication 1999:4, Lund, pp. 127–142.
[95] Ekvall, G., (1991), The organisational culture of idea–management: a creative climate for the management of ideas', in J. Henry, D. Walker (eds.), *Managing Innovations*, Sage Publications, London, pp. 73–79.
[96] Elkington J., T. Burke, (1990), *Groen zaken doen*, Maarten Muntinga, Amsterdam.
[97] Ellerman, A.D., R. Schmalensee, P.L. Jaskow, J.P. Montero, E.M. Bailey, (1997), *Emission Trading Under The U.S. Acid Rain Programme*, MIT Centre for Energy and Environmental Policy Research, Cambridge/Massachusetts, mimeo.
[98] European Environmental Agency (EEA) (1997) *Environmental Agreements, Case Studies*, Copenhagen, mimeo.
[99] Faber, A, D. van Welle, *Research and Development voor Ecologische Transities*, RMNO, Rijswijk, 2004.

[100] Fabrycki, W.S., B.S. Blanchard, (1991), *Life Cycle Cost and Economic Analysis*, Prentice Hall Inc., New Jersey.

[101] Fallen, C., (1983) Survival of the fittest technologies, *New Scientist*, No. 1859, 6 February p. 35–39.

[102] Fleischner, M, Key Issue related to the Legislation of Chemicals in the EU, in F. Leone, J. Hemmelskamp, (eds), *The Impact of EU – Regulation on Innovation of European Industry*, Expert meeting on Regulation an Innovation, Sevilla, 18–19 January 1989

[103] Flemming, D., (1996), Beyond the technical fix, in R. Welford en R. Starkey (eds.), *Business and the Environment*, Earthscan Publications, London, pp. 147–150.

[104] Foss, K., (1996), Transaction costs and technological development: the case of the Danish fruit and vegetable industry, *Research Policy*, No. 25, p. 531–547.

[105] Ganguly, A., (1999), *Business–driven Research and Development*, MacMillan Business, New York.

[106] Gatersleben, B., Ch. Vlek, (1988), Household Consumption, Quality of Life and Environmental Impacts: A Psychological Perspective and Empirical Study, in K.J. Noorman, T. Schoot Uiterkamp, *Green Households? Domestic Consumers, Environment and Sustainability*, Earthscan Publications, London, pp. 141–183.

[107] Georg, S., I. Ropke, U. Jorgensen, (1992), Clean Technology–Innovation and Environmental Regulation, *Environmental and Resource Economics*, Vol. 2: 533–550.

[108] Glachant, M., (2000), *Lessons from Implementation Studies: The Need of Adaptability in EU Environmental Policy Design and Implementation*, Working Paper, Ecole du Mines and European University Institute, Paris, mimeo.

[109] Goedkoop, M.J., C.J.G. van Halen, H.R.M. ter Riele, P.J.M. Rommens, (1999), *Product Service Systems*, mimeo.

[110] Gold, B., (1975), Technology, Productivity and Economic Analysis, in B. Gold (ed.), *Technological Change: Economics, Management and Environment*, Pergamon International Library, Pergamon Press, Oxford, pp. 1–42.

[111] Gallop, F.M., M.J. Roberts, (1983), Environmental Regulation and Productivity Growth: The Case of Fossil–fueled Electric Power Generation, *Journal of Political Economy*, Vol. 91, No. 4: 654–674.

[112] Graedel T.E., B.R. Allenby, (1995), *Industrial Ecology*, Prentice Hall, New Jersey.

[113] Grand, L. le, R. Robinson, (1976), *The Economics of Social Problems*, McMillan Press, London.

[114] Gray, B.W., R.J. Shadbegian, (1998), Environmental Regulation, Investment Timing, and Technology Choice, *The Journal of Industrial Economics*, Vol. XLVI: 235–256

[115] Gregori, T.R. de, (1987), Resources Are Not; They Become: An Institutional Theory', *Journal of Economic Issues*, Vol. XXI: 1241–1263.

[116] Griliches, Z., (1996), The Discovery of the Residual: A Historical Note, *Journal of Economic Literature*, Vol. XXXIV: 1324–1330.

[117] Griliches, Z., (1994), Productivity, R&D data and the Data Constraints, *American Economic Review*, March 1994, Vol. 84: 1–23.

[118] Gunningham, N., P. Grabowsky, (1998), *Smart Regulations: Designing Environmental Policy*, Oxford University Press, Oxford.

[119] Haq, G., P.D. Bailey, M.J. Chadwick, J. Forrester, J. Kuylenstierna, G. Leach, D. Villagrasa, M. Ferguson, I. Skinner, S. Oberthur, (2001), Determining the costs to industry of environmental regulation, *European Environment*, 11: 125–139.

[120] Hartje, V.J., (1984), (1984), *Adopting Rules for Pollution Control*, International Institut für Umwelt und Gesellschaft, Wissenschaftzentrum, Berlin, nr. 6.

[121] Hartje V.J., R.L. Lurie, (1984), *Adopting Rules for Pollution Control Innovations: End–of–Pipe versus Process Integrated Technology*, International Institut für Umwelt und Gesellschaft, Wissenschaftzentrum, Berlin.

[122] Hartog, H. den, R.J.M. Maas, (1990), Een duurzame economische ontwikkeling: macro–economische aspecten van een prioriteit voor het milieu, in P. Nijkamp, H. Verbruggen (eds), *Het Nederlands milieu in de Europese ruimte*, Stenfert Kroese Uitgevers, Leiden/Antwerpen.

[123] Havenman, R.H., G.R. Christainsen, (1985), Environmental Regulation and Productivity Growth (reprint from Natural Resource Journal, Vol. 21, No. 3, July 1981), in H. Peskin, P.R. Portney, A.V. Kneese, (eds.), *Environmental Regulation and the U.S. Economy*, Johns Hopkins University Press, Baltimore, p. 55–75.

[124] Heel, H.P. van, J.L.A.Jansen, (1990), *Duurzaam: zo gezegd, zo gedaan*, Technische Universiteit Delft, mimeo.

[125] Heertje, A., (1973), *Economie en technische ontwikkeling*, H.E. Stenfert Kroese BV, Leiden.

[126] Heinelt, H., A.E. Töller, (2003), Sustainability, Innovation, Participation and EMAS, in H. Heinelt and R. Smith (eds.), *Sustainability, Innovation and Participatory Governance*, Ashgate Publishers Ltd, Aldershot, UK, pp. 11–22.

[127] Heinelt, H., B. Meinke, R. Smith, G. Terizakis, (2003), Introduction, in H. Heinelt and R. Smith (eds.), *Sustainability, Innovation and Participatory Governance*, Ashgate Publishers Ltd, Aldershot, UK, pp. 267–282.

[128] Heiskanen, E. M. Pantzar, (1997), Towards Sustainable Consumption: Two New Perspectives, *Journal of Consumer Policy*, 20: 409–442.

[129] Herman, R., S.A. Ardekanin, J.H. Ausubel, (1989), Dematerialization, in J.H. Ausubel, H.E. Sladovich (eds.), *Technology and Environment*, National Academy Press, Washington DC, pp. 50–69.

[130] Heslinga, D.C., (1995), *Reinigen en ontvetten met gehalogeneerde oplosmiddelen en waterige systemen – een vergelijkende studie*, KWS 2000 rapporten nr. O10, mimeo.

[131] Heijden, W. van de, (1999), Als bedrijf is Schiphol gebonden aan milieuverplichtingen, *Polytechnisch Weekblad*, 1, pp. 2.

[132] Heijnes, H.A.M., H.J. Jantzen, C.A.J.C. Sedee, F. Schelleman, K. van den Berg, A.W.Dilweg, F. van Woerden, J. Okkema, A. Nentjes, (1997), *Milieu–emissies: kiezen voor winst!, Marktwerking in het milieubeleid: de potentiële kostenvoordelen van een systeem van Verhandelbare Emissierechten*, Interprovinciaal Overleg (IPO), mimeo.

[133] Heijungs, R., (1994), A generic method for the identification of options for cleaner products, *Ecological Economics*, No. 10: 69–81.

[134] Heyes, A., C. Liston–Heyes, (1997), Regulatory "Balancing" and the Efficiency of Green R&D, *Environmental and Resources Economics* No. 9: 493–507.

[135] Hoevenagel, R., U. van Rijn, L. Steg, H. de Wit, (1996), *Milieurelevant consumentengedrag, Ontwikkeling conceptueel model*, Sociaal Cultureel Planbureau, Rijswijk, pp. 19–62.

[136] Hofman, P.S., G.J.I. Schrama, (1999), *Innovations in the Dutch environmental policy for the industry target group*, paper for the eighth Greening of Industry Network Conference, Chapel Hill, USA, mimeo.

[137] Hofman, P.S., (2001), *Innovation, Negotiation and Path Dependencies in Industry and Policy*, paper for the ninth Greening of Industry Network Conference, Bangkok.

[138] Honig, E., A. Hanemaaijer, R. Engelen, A. Dekkers, R. Thomas, (2001) *Techno 2000; Modellering van de daling van eenheidskosten van technologieën in de tijd*, RIVM, rapportnummer 773008003.

[139] Hueting, R., (1990), *Correcting of National Income for Environmental Losses: a practical solution for a theoretical problem*, paper prepared for the Conference on Ecological Economics, Washington DC.

[140] Huisingh, D.L., M.H. Higler, N. Seldman, (1985), *Proven Profits from pollution prevention, Institute for Local Selfreliance*, North Caroline State University.

[141] Hyvättinen, H., Hildén, M., (2004), Environmental policies and marine engines—effects on the development and adoption of innovations, *Marine Policy*, Vol. 28: 491–502.

[142] Ianuzzi, A., (2002), *Industry Self-regulation and Voluntary Environmental Compliance*, Lewis Publishers, Seattle. Washington.

[143] Imkamp, H., (2000), The Interest of Consumers in Ecological Product Information Is Growing – Evidence From Two German Surveys, *Journal of Consumer Policy*, 23: 193–202.

[144] Jaffe, A.B., S.P. Peterson, P.R. Porteny, R.N. Stavins, (1995), Environmental Regulation and the Competitiveness of U.S. Manufacturing: What Does the Evidence Tell Us, *Journal of Economic Literature*, Vol. XXXIII: March: 132–163.

[145] Jaffe, A.B., K. Palmer, (1997), Environmental Regulation and Innovation: a panel data study, *The Review of Economics and Statistics*, Vol. LXXIX, No. 4, November: 610–619.

[146] Jaffe, A.B., R.G. Newell, R.N. Stavins, (2002), Environmental Policy and Technological Change, *Environmental and Resource Economics* 22: 41–69.

[147] Jaffe A.B., R.G. Newell, R.N., Stavins, (2005), The tale of two market failures: Technology and environmental policy, *Ecological Economics*, Vol. 54, Issue 2–3: 164–174.

[148] Jänicke, M., H. Monch, T. Ranneberg, (1986), *Umweltentlastung durch Structurwandel – eine Vorstudien für über 31 Industrieländer*, IIUG Berlin, Nr. 1, mimeo.

[149] Jänicke, M., M. Binder, H. Monch, (1997), *"Dirty Industries": Pattern of Change in Industrial Countries*, Environmental and Resource Economics, No. 9, p. 467–491.

[150] Jantzen, J., (1992), *Model on Sustainable Environmental Strategies (Moses)* (and data files on disc), Institute for Applied Environmental Economics, The Hague, mimeo.

[151] Jantzen, J., (red.), H. Heijnes, P. van Duijse (m.m.v. J–M Visser, M. Buist, B. van Diepen), (1995), *Technische vooruitgang en milieukosten*, Instituut voor Toegepaste Milieueconomie, mimeo.

[152] Jensen A.A., A. Remmen. (2004), *UNEP Guide to Life Cycle Management: A bridge to sustainable products*. UNEP, Paris.

[153] Jesinghaus, J., (2000), *Sustainability indicators in European Union*. World Conference of the International Political Science Association, Quebec, mimeo.

[154] Johansson, A., P. Kisch, M. Mirata, (2004), Distributed economies – A new engine for innovation, *Journal of Cleaner Production*, 13: 971–979.

[155] Jorna R.J., N.R. Faber, (2006), Sustainability: from environment and technology to people and organisations, in R. Jorna (ed.), *Sustainable Innovation, Greenleaf Publishing*, Sheffield UK, pp 28–41

[156] Kaarrer–Rueedi, E.E., (1996), Trends in Toxic Chemical Releases: A Case Study on the Drug Industry in Connecticut and New Jersey, *Journal of Environmental Planning and Management*, Vol. 39, No. 4: 577–592.

[157] Kanazawa, M.T., (1994), Water Subsidies, Water Transfers and Economic Efficiency, *Contemporary Economic Policy*, Vol. XII, April: 112–121.

[158] Karamanos, P., (2001), Voluntary Environmental Agreements: Evolution and Definition of a New Environmental Policy Approach, *Journal of Environmental Planning and Management*, Vol. 44, No. 1: 67–84.

[159] Karl, M., A. Moller, X. Matus, E. Grande, R. Kaiser, (2005), *Environmental Innovations: Institutional Impacts of Cooperation for Sustainable Development*, Fondazione Eni Enrico Matei, nota de lavoro 58.2005, mimeo.

[160] Kemp, R., X. Olsthoorn, F. Oosterhuis, H. Verbruggen, (1992), Supply and Demand Factors of Cleaner Technologies: Some Empirical Evidence, *Environmental and Resource Economics*, Vol. 2: 615–634.

[161] Kemp, R., (1995), *Environmental Policy and Technical Change*, MERIT, Universiteit van Maastricht (dissertation).

[162] Kemp, R., (1998), Environmental Regulation and Innovation Key Issues and Questions for Research', in F. Leone, J. Hemmelskamp (eds.), *The Impact of EU – Regulation on Innovation of European Industry*, Expert meeting on Regulation and Innovation, Sevilla, mimeo.

[163] Kemp, R., J. Schot, R. Hoogma, (1998), Regime Shifts to Sustainability Through Processes of Niche Formation: The Approach of Strategic Niche Management', *Technology Analysis & Strategic Management*, Vol. 10, No. 2: 175–195.

[164] Kemp, R. E. Moors, Modulating dynamics in transport for Climate Protection, in M. Faure, J. Gupta, A. Nentjes, *Climate Change and the Kyoto Protocol, The Role of Institutions and Instruments to Control Climate Change*, Edward Elgar, Cheltenham UK, Northampton, USA, 2003, pp. 313–340.

[165] Kip P., J. Krozer, (1990), *Financiële instrumenten in het afvalstoffenbeleid, ervaringen en mogelijkheden*, Tebodin Consulting Engineers, juli 1990, mimeo.

[166] Kip Viscusi, W., M.J. Moore, (1993), Product Liability, Research and Development and Innovation, *Journal of Political Economy*, Vol. 101, No 1: 161–184.

[167] Kip Viscusi, W.J.M. Vernon, J.E. Harrington Jr., (1995), *Economics of regulation and antitrust*, The MIT Press (second edition), Cambridge/Massachusetts, pp. 89–93 and pp. 711–719

[168] Kivimaa, P., P. Mickwitz, (2004), Driving Forces for Environmentally Sounder Innovations: the case of Finish Pulp and Paper Industry, in K. Jacob, M. Binder, A. Wieczorek (eds.), 2004, *Governance for Industrial Transformation*, Proceedings of the 2003 Berlin Conference on Human Dimension of Global Environmental Change, Environmental Policy Research Centre: Berlin, pp. 356–372.

[169] Klaassen, G.A.J., A. Nentjes, (1986), Macro–economische gevolgen van een intensieve bestrijding van zure regen, *Economische Statistische Berichten*, 12 december: 165–169.

[170] Klaassen, G.A.J., J.B. Opschoor, (1991), Economics of sustainability or the sustainability of economics: different paradigms, *Ecological Economics*, No. 4: 93–115.

[171] Klaassen, G.A.J., A. Nentjes, (1997), Sulfur Trading Under the 1990 CAAA in the US: An Assessment of First Experiences, *Journal of Institutional and Theoretical Economics*, Vol. 153: 384–410.

[172] Klink, J., J. Krozer, A. Nentjes, (1991), *Technologische ontwikkeling en economische instrumenten in het milieubeleid*, Nederlandse Organisatie voor Technologisch Aspectenonderzoek, Den Haag, p. 42–45.

[173] Klok, P–J., *Convenanten als instrument van milieubeleid*, Faculteit der Bestuurskunde, Universiteit Twente, Enschede, 1989 (disseration).

[174] Klomp, L., J.J.M. Pronk, (1998), *Kennis en economie 1998*, Centraal Bureau voor de Statistiek, Voorburg, pp. 50–64.

[175] Kloppenburg, G–J., (2000), *Vergunningverlening en het stimuleren van milieu–innovaties*, provincie Friesland, Leeuwarden, mimeo.

[176] Kneese, A.V., R.U. Ayres, R.C. d'Arge, (1970), *Economics and the Environment: A Material Balance Approach*, Resources for the Future, Washington DC.

[177] Konar, S., M.A. Cohen, (1997), Information As Regulation: The Effect of Community Right to Know Laws on Toxic Emissions, *Journal of Environmental Economics and Management*, No. 32: 109–124.

[178] Koten–Vermeulen, J.E.M. van, C.A. van der Heijden, F.X.R. van Leeuwen (1986), *Criteriadocument Fenol*, RIVM, rapportnr. 738513002, mimeo.

[179] Kothuis, B., J. Krozer, R. Kuil, (1997), *Environmental Assessment of solvent and water–based inks*, Institute for Applied Environmental Economics, The Hague, mimeo.

[180] Kremers, G.J., H.S. Buijtenhek, C. van Driel, J. Krozer, (1991), *Financieel instrumentarium voor regulering van verf met organische oplosmiddelen*, Tauw, Deventer, mimeo.

[181] Krozer, J., A. Nentjes, (1988), Economische infrastructuur en milieu, *Economische Statistische Berichten*, nr. 6 januari:. 34–38.

[182] Krozer, J., (1989), Vernieuwingen in milieutechnologie, in H. Vollenbergh (red.), *Milieu en Innovatie*, Wolters–Noordhoff, Groningen, pp. 117–134.

[183] Krozer, J., H.A. Nijenhuis, (1990), DESC: Decision model on Environmental Strategies of Corporations, in O.J. van Gerwen (red.), *Financiering van het milieubeleid*, Raad voor het Milieu– en Natuuronderzoek, december, pp. 101–105.

[184] Krozer, J., (1992), Decision model for Environmental Strategies of Corporations, in L. Preisner (ed.), *Institutions and Environmental Protection*, European Association of Environmental and Resource Economists, proceedings of the conference, Cracow Academy of Economics, Cracow.

[185] Krozer, J., (1992), *Milieukosten van OMO*, Tebodin Consulting Engineers, Den Haag, mimeo.

[186] Krozer, J., (1993), *Milieukosten van twee margarines*, Tebodin Consulting Engineers, Den Haag, mimeo.

[187] Krozer, J., M. Lavrano, (1994), *Kosten van tomaten*, Instituut voor Toegepaste Milieueconomie, Den Haag, mimeo.

[188] Duijse van P., J. Krozer, J, (1994), Costs and Benefits of Company's Environmental Strategies, Instituut voor Toegepaste Milieueconomie, Den Haag, mimeo.

[189] Krozer, J., (1994), *Ecodesign bij Ahrend*, Instituut voor Toegepaste Milieueconomie, Den Haag, mimeo.

[190] Krozer, J., (1995), *Milieukosten van katoenteelt*, verkenning voor het PRIMA–project, Instituut voor Toegepaste Milieueconomie, Den Haag, mimeo.

[191] Krozer, J., J. Cramer, P. Doelman, F. Schelleman, (1995), *Goede verpakking van een lastige boodschap*, rapport van voorbereiding van de Strategische Conferentie Verpakkingen, Instituut voor Toegepaste Milieueconomie, Den Haag, mimeo.

[192] Krozer, J., (1995), Milieurendement en markt, in J.J. Bouma, J.M.D. Kosten, H.R.J. Vollenbergh (red.), *Milieurendement in theorie en praktijk*, Samson HD Tjeenk Willink, Alphen aan de Rijn.

[193] Krozer, J., (1996), *Environmental Life Cycle Costing*, paper presented at the Roundtable Conference on Cleaner Production, Rotterdam.

[194] Krozer, J., J.C. Vis, (1998), How to get LCA in right direction?, *Journal of Cleaner Technologies*, Vol. 6, No. 1: 53–61.

[195] Krozer, Y. (2002), *Milieu en Innovatie*, Rijksuniversiteit Groningen (dissertation),

[196] Krozer, J. K. Maas, B. Kothuis, (2003), Demonstration of environmentally sound and cost–effective shipping, *Journal of Cleaner Production*, 11: 767–777.

[197] Krozer, Y., (2004), Social Demands in Life Cycle Management, *Greening Management International*, 45, Spring 2004: 95–106.

[198] Krozer Y, A.Nentjes, (2006), Environmental Policy and Innovations, *Business Strategies and the Environment*, published on line DOI 10.1002/bse.

[199] Krozer, Y. A. Nentjes, Milieu–innovatie en kosten van emissionreductie, Economische Statistische Berichten 27Jrg. 92 4515: 452–456.

[200] Kuil, R.E., J. Krozer, (1996) *IMAGE*, Instituut voor Toegepaste Milieu–economie, mimeo.

[201] Kuipers, S.K., A. Nentjes, (1973), Pollution in a neo–classical world: the classics rehabilitated?, *The Economist*, Vol. 121, No. 1: 52–67.

[202] Lafont, J–J., J. Tirole, (1994), Environmental policy, compliance and innovation, *European Economic Review*, Vol. 38: 555–562.

[203] Lanjouw, J.O., A. Mody, (1996), Innovation and the international diffusion of environmentally responsive technology, *Research Policy*, No. 25: 549–571.

[204] Larson, E.D., M.H. Ross, R.H. Williams, (1986), Beyond the era of materials, *Scientific American*, Vol. 34, No. 6: 24–29.

[205] Lassman, G., (1958), *Die Produktionsfunktion und ihre Bedeutung für die betriebswirtschaftliche Kostentheorie*, Westdeutscher Verlag, Köln/Opladen.

[206] Leeuwen, C. van, (1989), De organisatie van milieu en veiligheid in een grote onderneming, in H. Vollebergh (red.), *Milieu en Innovatie*, Wolters Noordhoff, Groningen, pp. 155–176.

[207] Leipert, C., (1986), Social Costs of Economic Growth, *Journal of Economic Issues*, Vol. XX, No. 1: 109–131.

[208] Lilley, S., (1980) Men, Machine and History, a short history of tools and machines in relation to social progress, Retrospect and Summary, in E.J. Fisher (red.), *Geschiedenis van de techniek*, Martinus Nijhoff, Den Haag, pp. 61–93.

[209] Lombardini, S., Economy versus Ecology, in F. Archibugi and P. Nijkamp (eds.), *Economy and Ecology: Towards Sustainable Development*, Kluwer Academic Publishers, Dordrecht, 1989, pp 139–147

[210] Magat, A., (1978), Pollution Control and Technological Advance: A Dynamic Model of the Firm, *Journal of Environmental Economics and Management*, No. 5: 1–25.

[211] Malecki, E.J., (191), *Technology and Economic Development*, Longman Scientific & Technical, London

[212] Malueg, D.A., (1989), Emission Credit Trading and the Incentive to Adopt New Pollution Abatement Technology, *Journal of Environmental Economics and Management*, No. 16: 52–57.

[213] Mansfield, E., (1971) Technical Change and the Rate of Imitation, in N. Rosenberg (ed.), *The Economics of Technological Change*, Penguin Modern Economics, UK, pp. 284–315.

[214] March, J.G., (1989) The technology of foolishness, in J.G. March (ed.), *Decision and Organisation*, Basil Blackwell, Oxford, pp. 253–265.

[215] Marin, A., Firm Incentive to Promote Technological Change in Pollution Control: Comment, *Journal of Environmental Economics and Management*, No. 21, 1991, 297–300.

[216] Martino, J.O. (1975), *Technological Forecasting for Decision makers*, American Elservier, New York.

[217] McCain, R.A., (1978), Endogenous Bias in Technical Progress and Environmental Policy, American Economic Review, Vol. 68, No. 4: 538–546.

[218] McConnel, V.D., G.E. Schwarz, (1993), The Supply and Demand for Pollution Control: Evidence from Wastewater Treatment, *Journal of Environmental Economics and Management*, No. 23: 54–77.

[219] Meadows, H., D.L. Meadows, J. Randers, (1991), *Beyond the Limits. Confronting Global Collapse; Envisioning a Sustainable Future*, Earthscan.

[220] Meer, J. van der, (1997), Soepele afspraken zonder ambities, *Intermediair*, jrg. 33, nr. 47: 23–25.

[221] Mendelsohn, R., (1984), Endogenous Technical Change and Environmental Regulation, *Journal of Environmental Economics and Management*, No. 1: 202–207.

[222] Mensink, A.J., C.H.A. Quarles van Ufford, J.M.M. Veeken, (1988), Naar een preventief milieubeleid: onderzoek naar belemmeringen en mogelijkheden van preventie bij de houtconserverings– en galvanische industrie, Nijmeegse Milieukundige Studies nr.1, Katholieke Universiteit Nijmegen, mimeo.

[223] Metclafe, J.S., (1994), Evolutionary Economics and Technology Policy, *The Economic Journal*, No. 104: 931–944.

[224] Meulen, A. van der, P.J. Rombout, C.J. Prins, (1987), *Criteriadocument Fijn Stof*, RIVM, rapportnr. 738513006, mimeo.

[225] Mill, J.S., (1985), *The Principles of Political Economy*, Book IV, Chapter VI, Penguin Classics, p. 116.

[226] Milliman, S.R., R. Prince, (1989), Firm Incentives to Promote Technological change in Pollution Control, *Journal of Environmental Economics and Management*, No. 17: 247–265.

[227] Ministerie van Volkshuisvesting, Ruimtelijke Ordening en Milieubeheer/VROM (1988), *Notitie bestrijding verzurende uitworp*, Tweede Kamer, kamerstuk 18 225.

[228] Mishan, E.J., (1993), *The Cost of Economic Growth*, Weidenfeld and Nicholson, London, pp. 30–46.

[229] Mood, A.M., F.A. Graybill, (1963), *Introduction to the Theory of Statistics*, McGraw–Hill Book Company, New York, pp. 70–73.

[230] Murdoch, J.C., T. Sandler, (1997), The voluntary provision of a pure public good: The case of reduced CFC emissions and the Montreal Protocol, *Journal of Public Economics*, No. 63: 331–349.

[231] Nakicenovic, N., (1991), Diffusion of Pervasive Systems: A Case of Transport Infrastructures, in N. Nakicenovic, A. Grubler (eds.), *Diffusion of Technologies and Social Behaviour*, International Institute for Applied System Analysis, Laxenburg, Austria, pp. 483–510.

[232] Nakicenovic, N., (2002), Technological Change and Diffusion as a Learning Process, in A. Grübler, N. Nakicenovic, W.D. Nordhaus (eds.), *Technological Change and the Environment*, Resources for the Future, pp. 127–159.

[233] Nelson, R, S. Winter, (1982), *An Evolutionary Theory of Economic Change*, Cambrigde MA, The Belknap Press of the Harvard University Press.

[234] Nelson, R., (1995), Recent Evolutionary Theorizing About Economic Change, *Journal of Economic Literature*, Vol. XXXIII, March: 48–90.

[235] Nentjes, A., D. Wiersma, (1987), Innovation and pollution control, *International Journal of Social Economics*, Vol. 15, No. 3–4.

[236] Nentjes, A., (1988), Marktconform milieubeleid, *Economische Statistische Berichten*, 27 April: 401–405.

[237] Nentjes, A., (1988), *An Economic Model of Innovation in Pollution Control Technology*, paper presented at the Annual Meeting at AAERE, New York.

[238] Nentjes, A., Scholten, H., (1989), Een financiële impuls voor het milieu?, kanttekeningen bij het initiatief–wetsvoorstel Vermeend–Melkert', *Tijdschrift voor Politieke Economie*, jrg. 12, nr. 1: 17–27.

[239] Nentjes, A., (1990), *Groei en bloei: economie en milieukwaliteit*, in Commissie Lange Termijn Beleid, Milieudenkbeelden voor de twintigste eeuw, Kerckebosch BV, Zeist, pp. 477–499.

[240] Nentjes, A., J. de Vries, (1990), *Financiële Instrumenten voor het Nederlands Milieubeleid*, Landelijk Milieu Overleg, Utrecht.

[241] Noci, G., R. Verganti, (1999), Managing "green" product innovation in small firms', *R&D Management*, Vol. 29, No. 1: 3–15.

[242] OTA, Office of Technology Assessment, Industry, (1994) *Technology, Environment, Competitive Challenges and Business Opportunities*, US Congress, Washington DC.

[243] Opschoor, J.B., H.B. Vos, (1989), *Economic instruments for environmental protection*, Organisation for Economic Co–Operation and Development, Paris.

[244] Ostman, A., W.W. Pommerehne, L.P. Feld, A. Hart, (1997), Umweltgemeingüter, *Zeitschrift für Wirtschafts– und Sozialwissenschaften* (ZWS), Nr. 117: 107–144.

[245] Palmer, K., H. Sigman, (1997), The Cost of Reducing Municipal Solid Waste, *Journal of Environmental Economics and Management*, No. 33: 128–150.

[246] Pearce, D.W., R. K. Turner, (1990), *Economics of Natural Resources and the Environment*, Harvester Wheatsheaf, London.

[247] Peppel, R.A. van, (1995), *Naleving van milieurecht; Toepassing van beleidsinstrumenten op de Nederlandse verfindustrie*, Kluwer, Deventer.

[248] Persson, M., (1999). Dankzij moderne techniek kan traditioneel vliegtuig overboord, *Polytechnisch Weekblad*, 15 December.

[249] Perman, R., Yue, M., J. McGilvray, (2006), *Natural Resources & Environmental Economics*, Longman, New York.

[250] Perrings, C., (1991), Reserved rationality and the precautionary principle: technological change, time and uncertainty in environmental decision making', in R. Costanza (ed.), *Ecological Economics*, Columbia Press, New York, pp. 153–167.

[251] Petrovsky, H., (1994), *The Evolution of Useful Things*, Vintage Books, Random House, New York.

[252] Pieters, J.H.M., (1997), Subsidies *and the Environment: on how subsidies and tax incentives may affect production decision and the environment*, paper for the UN Fourth Expert Group Meeting on Financial Issues of Agenda 21, Santiago, Chile.

[253] Pigou, A.C., (1920), *The Economics of Welfare*, MacMillan (first edition), London,

[254] Porter, M.E., C. van der Linde, (1995), *Green and Competitive: Ending the Stalemate*, Harvard Business Review, September/October 1995: 119–134.

[255] Porter, M.E., (1996), *Concurrentievoordeel*, Uitgeverij Contact (5^e druk), Amsterdam/Antwerpen.

[256] Quakernaat, J., J.A. Don, F. van den Akker, (1987), Process integrated environmental technology, a must to survive', in K.J.A. de Waal, W.J. van de Brink, (eds.), *Environmental Technology*, pp. 55–66.

[257] Quakernaat, J., J.A. Don, (1988), *Naar meer preventie–gerichte milieutechnologie in de industriële produktiesector*, Raad voor het Milieu– en Natuuronderzoek, nr. 27.

[258] Remner, D.S.; B.L. Low, G.T. Heaps–Nelson, (1994), *Air Pollution Control: Estimate the Cost of Scaleup*, Chemical Engineering, November, p. EE10–EE16

[259] Reijnders, L., (1984), *Pleidooi voor een duurzame relatie met het milieu*, Van Gennep, Amsterdam.

[260] Reijnders, L., (1996), *Environmentally Improved Production Processes and Products*, Kluwer Academic Publishers, Dordrecht/Boston/London.

[261] Rennings, K., A. Ziegler, K. Ankele, E. Hoffmann, (2006), The influence of different characteristics of the EU environmental management and auditing scheme on technical environmental innovation and economic performance, Ecological Economic, Vol. 57, Issue 1: 45–59.

[262] Riele, H.A. ter, A. Zweers, (1994), *Ecodesign: acht voorbeelden van milieugerichte produktontwikkeling*, TNO Produktcentrum in samenwerking met TU Delft faculteit Industrieel Ontwerpen.

[263] RIVM, Rijksinstituut voor Volksgezondheid en Milieuhygiëne, (1989), *Zorgen voor Morgen*, Samson H.D. Tjeenk Willink BV, Alphen aan den Rijn.

[264] Roome, N., (1994), Business Strategy, R&D Management and Environmental Imperatives', *R&D Management*, Vol. 24, No. 1:65–82.

[265] Ros, J.P.M., J. van der Plaat, (1986), Kosten voor de toepassing van terugwintechnieken in een galvanisch bedrijf, RIVM, rapportnr. 851403001, mimeo.

[266] Ros, J.P.M., W. Slooff, (1988), Integrated Criteria Document Cadmium, RIVM, Report No. 75847600, mimeo.

[267] Rose–Ackerman, S., (1983) Market models for Water Pollution Control, reprint in O'Riodan, R. Kerry Turner (eds.), Annotated Reader in Environmental Planning and Management, Pergamon Press, Oxford, pp. 63–86.

[268] Rosenberg, N., (1975), Problems in the economists conceptualization of technological innovation', in N. Rosenberg, Perspectives on Technology, Cambridge University Press, New York, pp. 61–84.

[269] Rosenberg, N., (1975), Technological innovation and natural resources: The niggardliness of nature reconsidered, in N. Rosenberg, Perspectives on Technology, Cambridge University Press, New York, pp. 229–259.

[270] Rosenberg, N., (1977), Innovative responses to material shortages, American Economic Review, Vol. 63, No. 2, p. 11–18, reprint in R. Dorfman, N.S. Dorfman, Economics of the Environment, W.W. Norton & Company Inc., New York, pp. 390–399.

[271] Rosenberg, N., (1982), The historiography of technical progress, in N. Rosenberg, Inside the black box, Cambridge University Press, New York, pp. 3–33.

[272] Rosenberg, N., (1982), Learning by using, in N. Rosenberg, Inside the black box, Cambridge University Press, New York, pp. 104–119.

[273] Rosenberg, N., L.E. Birdzell, (1986), How the West Grew Rich, I.B. Tauris & Co. Ltd, Publishers, London.

[274] Rosseger, G., (1980), The Economics of Production and Innovation, Pergamon Press, Oxford.

[275] Rothwell, R., (1992), Industrial innovation and government environmental regulation: Some lessons from the past, Technovation, Vol. 12, No. 7: 447–458.

[276] Rubik, F., G. Scholl, (2002), Integrated Product Policy in Europe–a development model and some impressions, Journal of Cleaner Production, 10 (2002): 507–515.

[277] Ruttan, V.W., (1971), Usher and Schumpeter on Invention, Innovation and Technological Change', in N. Rosenberg (ed.), The Economics of Technological Change, Penguin modern economics, New York, 1971, pp. 73–85.

[278] Ruttan, V.W., (1982), Agricultural Research Policy, University of Minnesota Press.

[279] Ruttan, V.W., (1997), Induced Innovation, Evolutionary Theory and Path Dependence: Sources of Technical Change, The Economic Journal, Vol. 107: 1520–1528.

[280] Ruttan, V.W. (2002), Source of Technical Change: Induced Innovation, Evolutionary Theory, and Path Dependency, in A. Grübler, N. Nakicenovic, W.D. Nordhaus (eds.), Technological Change and the Environment, Resources for the Future Washington DC, International Institute for Applied Systems, Laxenburg, pp.9–39

[281] Sarokin, D., W.R. Muir, C.G. Miller, S.R. Sperber, (1985), Cutting chemical waste, INFORM report, New York, mimeo.

[282] Saveage, D.E., A.L. White, (1994) New applications of Total Cost Assessment, Pollution Prevention Review, 3(3):247–259..

[283] Saviotti, P–P., (2005), On the Co–evolution of Technologies and Institutions, in M. Weber, J. Hemmelskamp, Towards Environmental Innovations Systems, Springer, Berlin, pp. 9–32.

[284] Schmidheiny, S, (1992), Changing the Course; A Global Business Perspective on Development and Environment, The MIT Press, Cambridge, Massachusetts.

[285] Schot, J., (1988), Regelgeving en technologische ontwikkeling', Tijdschrift voor Politieke Economie, jrg. 11, nr. 3: 79–96.

[286] Schot, J., (1991), *Maatschappelijke sturing van Technische Ontwikkeling*, Universiteit Twente, 1991 (dissertation).

[287] Schrama, G., (1987), Milieustrategie: inspelen op externe belanghebbenden, in K. Lulofs, G. Schrama (red.), *Ketenbeheer*, Jaarboek 1998, Centrum voor Schone Technologie en Milieubeleid, Twente University Press.

[288] Schultze, J., (1985), Economic Impact of chemicals control', *Chemistry and Industry*, 5:512.

[289] Schumpeter, J.A., (1989), *Business Cycles*, Porcupine Press, Philadelphia.

[290] Schuurman, J., (1988), *De prijs van water – een onderzoek naar de aard van de regulerende nevenwerking van de verontreinigingsheffing oppervlaktewater*, Gouda Quint BV, Arnhem, (disseration).

[291] Szejnwald–Brown, H., P. Vergragt, K. Green, L. Berchicci, (2003), Learning for Sustainability Transition through Bounded Socio–technical experiments in Personal Mobility, *Technology Analysis & Strategic Management*, Vol. 15, nr. 3: 291–315.

[292] Selden, T.M., D. Song, (1994), Environmental Quality and Development: Is There a Kuznets Curve for Air Pollution Emission, *Journal of Environmental Economics and Management*, No. 27: 147–162.

[293] Sheail, J., (1991), The regulation of Pesticides Use: An Historical Perspective, in L. Roberts, A. Weale (eds.), *Innovation and Environmental Risks*, Belhaven Press, London, pp. 38–46.

[294] Siebert, H., A.B. Antal, (1979), *The Political Economy of Environmental Protection*, JAI Press, Greenwich, Connecticut.

[295] Sigman, H.A., (1995), A comparison of public policies for lead recycling, *Rand Journal of Economics*, Vol. 26, No. 3: 452–478.

[296] Skillius, Å., U. Wennberg, (1998), *Continuity, Credibility and Comparability, key challenges for corporate environmental performance measurement and communication*, The International Institute for Industrial Environmental Economics, Lund University, Sweden, mimeo.

[297] Slooff, W., (1987), *Basisdocument benzeen*, RIVM, rapportnr. 758476001, mimeo.

[298] Slooff, W., (red.), (1987), *Ontwerp basisdocument propyleneoxide*, RIVM, rapportnr. 758473001, mimeo.

[299] Slooff, W., P.J. Blokzijl (red.), (1987), *Ontwerp basisdocument tolueen*, RIVM, rapportnr. 758473005, mimeo.

[300] Slooff, W., R.F.M.J. Cleven, J.A. Janus, J.P.M. Ros, (1987), *Ontwerp basisdocument koper*, RIVM, rapportnr. 758474003, mimeo.

[301] Slooff, W., H.C. Eerens, J.A. Janus and J.P.M. Ros, (1989), *Integrated Criteria Document Fluorides*, RIVM, Report No. 758474010 (a), mimeo.

[302] Slooff, W., J.A. Janus, A.J.C.M. Matthijsen, G.K. Montizaan, J.P.M. Ros (eds.), (1989) *Integrated Criteria Document PAHs*, RIVM, Report No. 758474011 (b) mimeo.

[303] Smith, A., (1986), On the natural progress of opulence, reprint in A. Smith, *The Wealth of Nations*, Book III, Chapter I, Penguin Classics, London, p. 481.

[304] Solow, R.M., (1977) The Economics of Resources or the Resources of Economics, reprint in R. Dorfman, N.S. Dorfman (eds.), *Economics of the Environment*, W.W. Norton & Company Inc. New York, pp. 354–370.

[305] Solow, R.M., (1974), Intergenerational Equity and Exhaustible Resources, *The Review of Economic Studies*, Symposium, p. 39–45.

[306] Stobaugh, (1988), *Innovation and Competition*, Harvard Business School Press, Boston.

[307] Sørup, P., (2000), *Technology, Innovation and Environment; Integrated Pollution Prevention and Control*. The European IPPC Bureau at Work, paper at Euro Environment, Aalborg.

[308] Sunnevåg, (1988), K., *Voluntary Agreements and the incentives for innovation*, paper for the CAVA workshop, Ghent, 26-27 November.

[309] Steen, B., S–O. Ryding, (1993) *The EPS Enviro–accounting Method*, Swedish Waste Research Council, mimeo.

[310] Stoneman, P., (1983), *The Economic Analysis of Technological Change*, Oxford University Press, Oxford.

[311] Stoneman, P., P. Diederen, (1984), Technology Diffusion and Public Policy', *The Economic Journal*, No. 104: 918–930.

[312] Straaten van der, J., (1994), Het Nederlands Milieubeleid in F. Dietz, W. Hafkamp, J. van der Straaten (eds.), *Basisboek Milieu–economie*, Boom, Amsterdam–Meppel.

[313] Thomson, P., T.G. Taylor, (1995), The Capital–Energy substitutability Debate: A New Look, *The Review of Economics and Statistics*, Vol. LXXVII: 564–569.

[314] Tietenberg, T.H., (1984), *Marketable Emission Permits in Principle and Practices*, paper at the Conference Economics of Energy and Environmental Policies: State of Art and Research Priorities, Yxtaholm, Sweden, mimeo.

[315] Tietenberg, T.H., (1991), Economic Instruments for Environmental Regulation', in D. Helm (ed.), *Economic Policy towards the Environment*, Blackwell, Oxford, pp 86–110.

[316] Tietenberg, T., (1994), *Environmental Economics and Policy*, HarperCollins College Publishers.

[317] Tilton, J.E., (1991), Material Substitution: The Role of New Technology, in N. Nakicenovic, A. Grubler (eds.), *Diffusion of Technologies and Social Behaviour*, Springer Verlag, Berlin, pp. 383–406.

[318] Tolsma, H., (2000), Strengere milieunormen jagen de vliegtuigtechnologie aan, *Polytechnisch Weekblad*, 5 januari.

[319] Tomer, J.F., (1992), The human firm in the natural environment: a socio–economic analysis of its behaviour', *Ecological Economics*, No. 6: 119–138.

[320] Tomer, J.F., T.R Sadler, (2007), Why we need a commitment approach to environmental policy, *Ecological Economics*, Volume 62, Issue 3–4: 627–636

[321] Tushman, M.L., P. Anderson, (1987), Technological Discontinuities and Organisation Environments, in A.M. Pettigrew (ed.), *The Management of Strategic Change*, Basil Blackwell, USA, pp. 89–122.

[322] Tuulenheimo, V., R. Thun, M. Backman, (1996), *Tools and Methods for Environmental Decision Making in Energy Production Companies*, IIIEE Communications, Lund.

[323] Uusitalo, L., (1983), Environmental impacts of changes in consumption style, in L. Uusitalo, *Consumer Behaviour and Environmental Quality*, Gower, Aldershot, UK, p. 123–142.

[324] Veering, A., (1993), *De verzekering dekt niet alle schade*, MilieuMarkt, oktober.

[325] Velthuizen, J.W., (1995)., *Determinants of investment in energy conservation*, Rijksuniversiteit Groningen, dissertation.

[326] Verbruggen, H., (1991), Milieutechnologie: de gestuurde deus ex machina, in J.W.A. van Dijk, L.G. Soete (red.), *Technologie in een open economie*, Samson, Bedrijfsinformatie, Alphen aan den Rijn/Zaventem.

[327] Verheul, H., P.J. Vergragt, (1995), Social Experiments in the Development of Environmental Technology: A Bottom–up Perspective, *Technology Analysis & Strategic Management*, Vol. 7, No. 3.

[328] Vickers, I., M. Cirdney–Hayes, (1999), Cleaner Production and Organisational Learning', *Technology Analysis & Strategic Management*, Vol. 11, No. 1: 75–94.

[329] Voges, I., Veerkamp, W., (2002), Corporate needs in Environmental and Social Assessment. *IAIA Business and Industry Series*. No. 1: pp. 2.

[330] Vollenbergh, H., (2007), *Impact of Environmental Policy Instruments on Technical Change*, OECD, Environment Directorate, mimeo,

[331] Vögtlander, J. A. Bijma, H.C. Brezet, (2002), Communicating the eco–efficiency of product and services by means of the eco–cost/value model. *Journal of Cleaner Production*, Vol. 10. Issue 1: 57–67.

[332] Wagner, G.R., (1991), Entrepreneurship and Innovation from an Environmental Risk Perspective, in L. Roberts, A. Weale, *Innovation and Environmental Risk*, Belhave Press, London, pp. 138–148.

[333] Wallace, D., (1995), *Environmental Policy and Industrial Innovation*, Strategies in Europe, the US and Japan, Earthscan, London.

[334] Watanabe, Ch., Ch. Griffy–Brown, B. Zhu, A. Nagamatsu, (2002), Inter–firm Technology Spillover and the "Virtous Cycle" of Photovoltaic development in Japan, in A. Grübler, N. Nakicenovic, W.D. Nordhaus (eds.), *Technological Change and the Environment*, Resources for the Future, pp. 127–159.

[335] Weaver, P. L. Jansen, G. van Grootveld, E. van de Spiegel, P. Vergragt, (2000), *Sustainable Technology Development*, Greenleaf Publishers, Sheffield.

[336] Weidner, H., (1985), Von Japan Lernen? Erfolge und Grenzen einer technokratischen Umweltpolitik, in S. Tsuru, H. Weidner (eds.), *Ein Model für uns: Die Erfolge der japanischen Umweltpolitik*, Kiepenheuer & Witsch, pp. 179–213.

[337] Weimann, J., (1990), *Umwelt–ökonomik, eine theorieorientierte Einführung*, Springer Verlag, Berlin.

[338] Weitzman, M.L., (1977), Sustainability and Technical Progress', *Scandinavian Journal of Economics*, Vol. 99, No. 1: 1–13.

[339] Weizsäcker, E. von, A.B. Lovins, L. Hunter Lovins, (1998), *Factor Four*, Earthscan, London.

[340] Weterings, R.A.P.M., J.B. Opschoor, (1992), *Milieugebruiksruimte als uitdaging voor technologische ontwikkeling*, RMNO rapportnr. 74 (a), Rijswijk, mimeo.

[341] Weterings, R.A.P.M., J. Kuijper, E. Smeets, (1997), *81 Mogelijkheden, Technologie voor duurzame ontwikkeling*, TNO Studiecentrum voor Technologie en Beleid, mimeo.

[342] White, A.L., M. Becker, J. Goldstein, (1991), *Total Cost Assessment, Accelerating industrial Pollution Prevention through innovative project financial analysis*, Tellus Institute Boston, mimeo.

[343] Wiersma, D., (1989), *De Efficiëntie van een Marktconform Milieubeleid, een uitwerking van de SO₂ emissiebestrijding van de Nederlandse Electriciteitssector*, Rijksuniversiteit Groningen, dissertation.

[344] Winsemius, P., (1986), *Gast in eigen huis*, Wolters/Samson, Alphen aan de Rijn.

[345] Winter, G., (1987), *Das Umweltbewusste Unternehmen*, BAUM AG, München.

[346] Wit, R.C.N., B.A. Leurs, G. de Wit, (1999), *Kosten en baten van milieuconvenanten in vergelijking met marktconforme instrumenten*, Centrum voor Energiebesparing en Schone Technologie, Delft, mimeo.

[347] Withagen, C., (1999), *De Porter hypothese: een verkenning van literatuur*, Raad voor het Milieu– en Natuuronderzoek, Rijswijk.

[348] Woerd, F. van der, (1997), *Self–regulation in corporate environmental management: changing interactions between corporations and authorities*, dissertation Vrije Universiteit.

[349] WCED, World Commission on Environment and Development, (1987), *Our Common Future*, Oxford University Press, Oxford, p. 43.

[350] Wright, G., (1997), Towards a more historical approach to technological change', *The Economic Journal*, Vol. 107: 1560–1566.

[351] Yaisawarng, S., J.D. Klein, (1994), The effects of sulfur dioxide controls on productivity change in the US Electric Power Industry, *The Review of Economics and Statistics*, Vol. LXXVI: 447–460.

[352] Zimmerman, K., (1985), *Präventive Umweltpolitik und technologische Anpassung*, International Institut für Umwelt und Gesellschaft, Wissenschaftzentrum, Berlin.

Index